MICROBIOLOGY RESEARCH ADVANCES

MICROCOSMS

ECOLOGY, BIOLOGICAL IMPLICATIONS AND ENVIRONMENTAL IMPACT

Microbiology Research Advances

MICROBIOLOGY RESEARCH ADVANCES

MICROCOSMS

ECOLOGY, BIOLOGICAL IMPLICATIONS AND ENVIRONMENTAL IMPACT

CHRISTOPHER C. HARRIS
EDITOR

nova publishers
New York

For permission to use material from this book please contact us:
Telephone 631-231-7269; Fax 631-231-8175
Web Site: http://www.novapublishers.com

Library of Congress Cataloging-in-Publication Data

ISBN: 978-1-62618-661-3

Library of Congress Control Number: 2013935915

Published by Nova Science Publishers, Inc. † New York

CONTENTS

PREFACE

In this book, the authors discuss the ecology, biological implications and environmental impact of microcosms. Topics include using microcosms to investigate aspects of plant-bacterial interactions and bacterial evolution; multifactorial microcosm experiments to predict how plant species and assemblages respond to changes in the availability and spatio-temporal heterogeneity of resources like water, light and nutrients; development and optimization of an aquatic laboratory microcosm for ecotoxicological risk assessment; assessment on transformation of organic pollutants in microcosms; bacterial evolution in simple microcosms; ecology and environmental side-effects of pesticides in tropical microcosms; floating dish microcosms to study the developments of biofilm communities; the role of semiosis and cohesion and sustainability inside microcosms; and soil microcosms and biogeographical research.

Chapter 1 – The fast pace of scientific research often means that contemporary investigators are unaware of critical developments resulting from the work of previous generations of scientists. The authors were superficially aware of Sergei Winogradsky (1856 – 1953) through his columns, currently in teaching to illustrate aspects of microbial succession and community function, but were unaware of his importance as the founder of microbial ecology and the first to use microcosms to study bacterial physiology. Here the authors take the opportunity to remind fellow microbiologists of his work, presenting in homage columns constructed using water and sediments from the Dnipro (Dnieper) collected close to his place of birth. They then provide a review of their own research using microcosms to investigate aspects of plant-bacterial interactions, bacterial evolution, biofilm-formation, and soil colonization, as well as recent advances in developing plastic and transparent soils to investigate root development and fungal hyphae invasion in soil pore networks, to show that more than one hundred years after the invention of the microcosm, they are still being used profitably in microbial research.

Chapter 2 – An important part of ecological research has been devoted to understand and predict how plant species and assemblages respond to changes in the availability and spatio-temporal heterogeneity of resources like water, light and nutrients. In the natural world, spatial heterogeneity in the availability of soil-based resources at various scales (soil heterogeneity) is the norm, rather than the exception, in most ecosystems. Albeit soil heterogeneity is a key feature of most terrestrial ecosystems, and plays a crucial role in shaping their structure, productivity, composition and diversity, it has often been ignored when evaluating the effects of global change drivers on plant performance and associated ecosystem processes. Thus, it is virtually unknown whether observed plant responses to global change drivers are modified by soil heterogeneity. This chapter synthesizes a series of multifactorial microcosm experiments aiming to evaluate the joint effects of soil heterogeneity and global change drivers (biotic diversity, nutrient availability, [CO_2] and changes in rainfall patterns) on the response of individuals and assemblages (in terms of biomass production and allocation and nutrient uptake patterns) formed by different combinations of *Lolium perenne* L., *Plantago lanceolata* L., *Trifolium repens* L., *Anthoxanthum odoratum* L. and *Holcus lanatus* L. The authors tested the hypothesis that soil heterogeneity modulates the magnitude of the response of both individuals and assemblages to multiple global change drivers (i.e., that individual and assemblage responses to soil heterogeneity and global change drivers will not be predictable from the observed responses to any one of the evaluated factors in isolation). In all the experiments conducted the authors did find a series of two- and three-term interactions determining key responses at both the individual and assemblage level. Thus, the authors' working hypothesis was supported by their results. Their synthesis strongly suggest that single-factor studies should not be used to extrapolate how grassland ecosystems may respond to global change, which is composed of multiple—and potentially interacting—drivers.

Chapter 3 – Ecotoxicological assessment of chemical substances is mainly based on standard bioassays, most often single-species tests covering the range of acute and chronic toxicity. In the authors' laboratory the authors have been developping since 1997 a protocol of ecotoxicological bioassay in 2-L laboratory microcosms and applied it to the study of various pollutants and scenarios of ecotoxicological risk assessment in the field of urban facilities and transport infrastructures. Effects are assessed on five different organisms (micro-algae, duckweeds, daphnids, amphipods, chironomids) using endpoints such as growth, emergence (chironomids), reproduction (daphnids), survival, with a duration exposure of 3-4 weeks. For specific studies, some variants of the 2-L protocol have been developed in higher volumes (up to 100 L) with increasing biological diversity and different hydraulic conditions. More recently, a flow-through microcosm assay (dynamic assay) was developped for the 2-L protocol to improve conditions inside the microcosms, resulting in stabilisation of physico-chemical parameters, increase of organisms fitness and reduction of variability.

This chapter will present the various protocols, their application to different ecotoxicological risk assessment studies, their limitations (variability and sensitivity, ecological relevance, organism fitness, ...), and the improvement of the 2-L protocol through continuous water renewal. Finally, recent works and perspectives in modelling will be presented.

Chapter 4 – Investigation on transformation of organic pollutants in microcosm is an important way of understanding their fate and behavior in the environment. The common methods used in microcosm experiment include (1) determination of concentration change of organic pollutants during the whole process by gas chromatography (GC) or high performance liquid chromatography (HPLC); (2) identification of degradation products by gas chromatography-mass spectrometry (GC-MS) or high performance liquid chromatography-mass spectrometry (HPLC-MS).

Over the past decades, a new and innovative method of compound-specific isotope analysis (CSIA) using gas chromatography coupled isotope ratio mass spectrometry via a combustion interface (GC-C-IRMS) has been established as a powerful tool for assessing transformation of organic pollutants based on the fractionation of stable isotopes during degradation process.

Two stable isotopes, ^{13}C and ^{2}H, have been extensively exploited to evaluate the transformation processes of contaminants. The enrichment of $\delta^{13}C$ and $\delta^{2}H$ offer a direct evidence of transformation, which does not prominently influenced by physical processes such as volatilization and sorption. Moreover, isotopic fractionation effect is closely associated with the reaction mechanism and the type of bond being broken. In this chapter, the authors review studies on degradation pathway of organic pollutants in microcosm by mean of CSIA. Especially with diethyl phthalate (DEP) as an example, the authors discuss how the CSIA technique combined with GC-MS analysis was used for studying the degradation pathway of organic pollutants in microcosm.

Chapter 5 – Microcosms have been used extensively to investigate ecological factors, evolutionary dynamics and the underlying molecular biology of adapting populations of bacteria in experimental evolution studies. One such experimental system has combined simple glass vials containing King's B liquid growth medium with the model soil and plant-associated pseudomonad, *Pseudomonas fluorescens* SBW25. In these, evolution reproducibly generates the Wrinkly Spreader genotype, a novel class of adaptive mutations that enable *Pf.* SBW25 to produce a robust, physical cohesive–class biofilm at the air-liquid interface. The analysis of Wrinkly Spreader mini-Tn*5* mutants led to the identification of cellulose as the main matrix component of the biofilm, and ultimately to the discovery of the mutation in a chemosensory-like system that activated the Wrinkly Spreader phenotype through the over-production of cyclic-*di*-GMP. Parallel ecological investigations have shown the importance of O_2 availability for the evolution and success of the Wrinkly Spreader. The initial colonisers establish an O_2 gradient, dividing the microcosm into a 100 – 200 μm high-O_2 zone at the air-liquid interface, and second, deeper low-O_2 zone in the rest of the liquid column. Growth is O_2–limited in King's B medium, and it is the domination of the air-liquid interface that guarantees access to O_2 and underlies the fitness advantage the Wrinkly Spreader has over non-biofilm–forming competitors including the ancestral wild-type *Pf.* SBW25. However, recent experiments demonstrate that Wrinkly Spreader fitness is reduced in static microcosms containing lower levels of nutrients, or in those in which King's B liquid viscosity has been increased. These findings demonstrate that the Wrinkly Spreader biofilm is a finely-balanced solution to the colonisation of the air-liquid interface in static microcosms, and changing environmental conditions will impact directly on Wrinkly Spreader fitness. This work highlights the importance of environmental factors, which along with competition, help guide the evolution of *Pf.* SBW25 in simple microcosms, invariably leading to the rise of the Wrinkly Spreaders.

Chapter 6 – It has often been discussed that research into the environmental fate and effects of pesticides has been almost exclusively conducted under non-tropical conditions. Furthermore, those studies conducted in tropical regions have focused almost exclusively on single species laboratory tests and very few microcosm studies under tropical settings are currently available. Given the differences in ecology and climatic conditions between tropical and non-tropical conditions, differences in pesticide fate, direct and indirect effects and recovery potential of impaired ecosystems may be anticipated. In the present chapter, findings from studies conducted in tropical Thailand are included to illustrate these possible differences. Indications for the methodology and research needs of future tropical microcosm studies are discussed.

Chapter 7 – Bacteria in the environment are organized in communities and show often high degrees of worksharing and cooperation. They are resilient against noxious agents and their functions are redundant to compensate changing environmental conditions. The authors' understanding of these functional communities, however, is limited, especially because only a minority of strains present in the environment can currently be isolated. Microcosms are powerful tools for the study of microbial communities without the need for prior isolation of strains. The authors became especially interested in microbial communities growing on interfaces forming so called biofilms. To study functional communities organized in biofilms they developed floating dish microcosms. Here soil or sediment samples from the environment are placed in a pot, covered with sterile water and a dish carrying different materials underneath is floated on the surface. Bacteria now migrate through the water column to these substrata and form complex biofilms. The authors used these biofilms to i) isolate degraders of pollutants used as substrata (polychlorinated biphenyls, hexachlorocyclohexanes, tributyltinoxide), ii) monitor biofilm development over time observing biofilm structure and composition and the authors identified the succession of degraders of different congeners, iii) compare biofilm communities from different sites and found similar activities but different species composition from different sites and iv) manipulate biofilm composition by using different substrata or adding compounds known to control biofilm formation. The floating dish microcosms can easily be adapted to specific research requirements making the whole approach very flexible. The authors optimized the protocol so that the microcosms can be used to grow biofilm communities which can be characterized by confocal laser microscopy, phylogenetic fingerprinting of the community composition and chemical analysis of pollutant degradation. The microcosms can also be adapted for the screening of natural compounds or novel materials for their ability to shape biofilm composition, structure and activity. In this chapter they present different floating dish microcosms and their scientific results. They also discuss the potential but also the limitations these microcosms have.

Chapter 8 – The dialectic relationship that characterises the semiotic process inherent to any form of cognition is responsible for the cohesion and sustainability of the microcosms defined by living entities in their interaction with the surrounding world.

The coupling that arises as a result of the plasticity of both living organism and its environmental bubble produces an autonomous and strictly bounded unit that has nevertheless been shaped extensively in the course of its evolutionary and developmental history.

Semiosis is consequently an embodied, embedded and always situated phenomenon. This means it involves a living entity endowed with a particular physical architecture interacting with the specific world it is immersed in, producing a dynamics that defines events taking place in a particular microcosm and perceived by a human observer as anchored in space/time.

Uexkull (1909) introduced the fundamental concepts of Umwelt and Innenwelt stressing their strict interdependent nature. The present paper stems from these fundamental concepts but it goes further by equating them with the semiosis inherent to any form of cognition. Plus, by proposing the mathematical modelling of a prototypical instance of individual cognition and by introducing the concept of Umwelt Overlap the paper attempts to identify the essential parameters of ecological sustainability.

In: Microcosms
Editor: Christopher C. Harris

ISBN: 978-1-62618-661-3

Chapter 1

FROM WINOGRADSKY'S COLUMN TO CONTEMPORARY RESEARCH USING BACTERIAL MICROCOSMS

Olena Moshynets[1], Mariia Boretska[2] and Andrew J. Spiers[3]*

[1]Laboratory of Microbial Ecology, Institute of Molecular Biology and Genetics, National Academy of Sciences of Ukraine, Kyiv, Ukraine

[2]Department of General and Soil Microbiology, D. K. Zabolotny Institute of Microbiology and Virology, National Academy of Sciences of Ukraine, Kyiv, Ukraine

[3]The SIMBIOS Centre and School of Contemporary Sciences, University of Abertay Dundee, Dundee, UK

ABSTRACT

The fast pace of scientific research often means that contemporary investigators are unaware of critical developments resulting from the work of previous generations of scientists. We were superficially aware of Sergei Winogradsky (1856 – 1953) through his columns, currently in teaching to illustrate aspects of microbial succession and community function, but were unaware of his importance as the founder of microbial ecology and the first to use microcosms to study bacterial physiology. Here we take the opportunity to remind fellow microbiologists of his work, presenting in homage columns constructed using water and sediments from the Dnipro (Dnieper) collected close to his place of birth. We then provide a review of our own research using microcosms to investigate aspects of plant-bacterial interactions, bacterial evolution, biofilm-formation, and soil colonization, as well as recent advances in developing plastic and transparent soils to investigate root development and fungal hyphae invasion in soil pore networks, to show that more than one hundred years after the invention of the microcosm, they are still being used profitably in microbial research.

* E-mail address: a.spiers@abertay.ac.uk.

INTRODUCTION

The science of medical microbiology was established through the pioneering work of scientists such as Louis Pasteur (1822 – 1895), Ferdinand Cohn (1828 – 1898), Heinrich Koch (1843 – 1910), and others in the last part of the 19th century. Pasteur made significant breakthroughs investigating the cause and prevention of diseases, Cohn instigated bacterial classification based on morphology, and Koch became famous for his postulates defining pathogens. During this period, the early isolation and characterisation of bacteria relied on the use of universal growth media and the presumption that all microbial activity was the result of specific isolates functioning alone. In contrast, the pioneering work of the environmental and ecological microbiologists, Martinus Beijerinck (1851 – 1931) and Sergei Winogradsky (1856 – 1953), are often overlooked and are perhaps unfairly overshadowed by their contemporaries who studied human disease. Beijerinck and Winogradsky investigated the biochemical role bacteria play in important natural processes, including nitrification, nitrogen fixation and sulfur[1] oxidation. Winogradsky emphasised repeatedly that the objectives of ecological microbiology differed from those of medical microbiology [109, 110]. He pioneered the use of the selective[2] approach to study biological processes in contrast to the universal medium used by Koch and others to isolate and characterise pure strains (see the quote from Winogradsky [105, 106] in [109]), as well as the realisation that in order to understand some biological processes, communities of bacteria needed to be studied rather than single cultures. Winogradsky is often described as the father of microbial ecology, and is considered the founder of contemporary ecological microbiology in Russia where his approach led to the Winogradsky School of Microbiology[3] [2, 17, 109, 110]. More relevant to this book is that Winogradsky was one of the first to use and promote microcosms in microbial research.

S. N. WINOGRADSKY

Sergei Nikolaevich Winogradsky[4] was born in 1856 into a wealthy family in Kyiv (Kiev), then part of Russia (a photograph of him is presented in Figure 1). He appeared to have had a troubled education, selling his golden graduation medal from the Second Kyiv Gymnasium (a secondary or grammar school) in 1873 in response to what he felt was an '*uninteresting, disagreeable, physically and morally humiliating style of teaching*' (Selam Waksman, quoted in [17]). He entered the University of Kyiv in 1873 to study law following his brother and father, but transferred to the natural sciences within a month before leaving the university altogether. He spent a year at the St. Petersburg Academy of Music studying piano, before finally entering the University of St. Petersburg in 1877 to study the natural sciences and microbial physiology with Andrey Famintsyn.

[1] We adopt the spelling convention for sulfur and sulfur compounds as recommended by the Royal Society of Chemistry Nomenclature Committee.
[2] Winogradsky referred to this as 'elective'.
[3] In comparison, Beijernck's Delft School of Microbiology was relatively well-known to non-Soviet microbiologists and historians of biology.
[4] Also transliterated from the Russian (Виноградский) or Ukrainian (Виноградський) as 'Vinogradskii'.

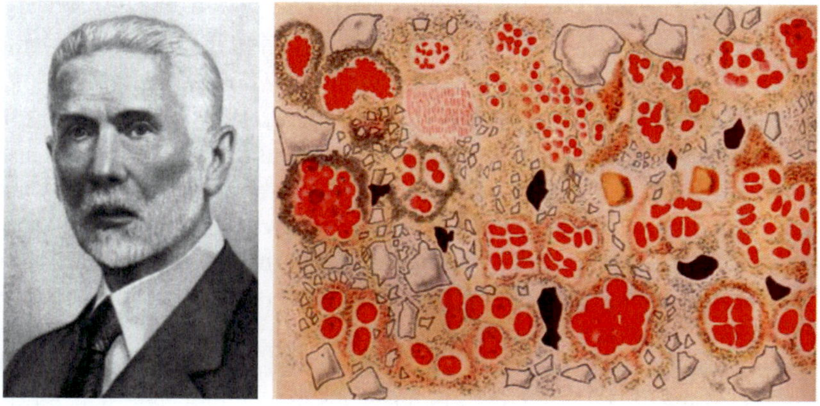

Figure 1. S.N. Winogradsky and soil landscapes. (Left) Photograph of Winogradsky (c. 1936). (Right) Winogradsky drew 'soil landscapes' to show the relationship between soil microorganisms and particles (Brie soil sample, ~100 μm wide; c. 1924). Images are from Winogradsky [105] and Seliber [82].

An outline of his impressive professional career is provided in the table at the end of this Chapter, and several relevant areas of his work are described in this section (the details summarised here are compiled from various references including [1, 2, 3, 17, 75, 78, 82, 104, 109, 110]; we do not cite Winogradsky's research publications here as they can be accessed through the references provided). In Famintsyn's laboratory, Winogradsky examined the nutritional and physiological properties of the yeast, *Candida vini*, originally known as *Mycoderma vini*[5]. *C. vini* is a pathogen of sugar beet (*Beta vulgaris*), which was causing problems for the developing European sugar industry, and it is also an etiological agent of wine illness or mould. Winogradsky used Geissler chambers to investigate the growth of *C. vini* cultures under different conditions. Detailed descriptions of this apparatus originally used by Pasteur [61] and its use by Winogradsky can be found in [2, 17]. In his biography of Winogradsky, Waksman suggests that these investigations of *C. vini* are the '*first careful investigations ever made on the influence of controlled environment on the growth of microorganisms in pure culture, under well-defined experimental conditions*' (Waksman, quoted in [17]). They may also be the first significant use of microcosms in the study of microorganisms, and Famintsyn's approach of coupling nutritional and physiological experiments with microscopic observations became Winogradsky's *modus operandi* [17]. Having completed his Master of Science degree at the University of St. Petersburg, Winogradsky joined Anton DeBary at the University of Strasburg in 1885 where he studied the filamentous sulfur-oxidizing bacterium *Beggiatoa*[6]. He was initially assigned a project to determine whether microorganisms were monomorphic or pleomorphic[7]. Winogradsky chose to investigate this using *Beggiatoa*, but could not prove the case that it was monomorphic as he was unable to isolate sufficiently pure cultures from sulfur springs in Switzerland and Germany. However, he became interested in the ability of *Beggiatoa* to accumulate intracellular sulfur granules and the role of sulfur metabolism in the growth of this bacterium.

[5] For a description of the *Candida* see [86].
[6] For a description of the *Beggiatoa* see [98].
[7] At the time there was a significant debate as to whether microorganisms exhibited multiple forms (i.e. pleomorphic) or whether they only had one form (i.e. monomorphic) (now the current view). However, to prove monomorphism, pure cultures where required for investigation.

During this period Winogradsky also '*reconfigured his technical repertoire* ... (adding) *new variations of slide microcultures and, most important, artificial environments*' to investigate bacteria in their natural state, allowing him to bring the '*field into his laboratory*' [2].

Winogradsky demonstrated that the *Beggiatoa* oxidized H_2S (hydrogen sulfide) to S^0 (sulfur) to release energy and formed the sulfur granules[8] as a result of that process, referring to this form of metabolism as 'inorgoxidation' (the oxidation of inorganic compounds). By observing the behaviour of *Beggiatoa* cells under a cover-slip, Winogradsky was able to demonstrate that the cells migrated to a narrow interface where both H_2S and O_2 (oxygen) co-exist, O_2 being required for the oxidation of H_2S. This was also an early demonstration that motile organisms will choose an optimal position along chemical gradients where the most efficient metabolism and rapid growth is possible. A corollary of this is that simple microcosms with two chemical gradients can provide a habitat choice to which organisms can respond. From his studies of the oxidation of H_2S by *Beggiatoa*, Winogradsky formulated the theory of chemolithotrophy in which inorganic substrates, usually of mineral origin, are used to obtain reducing equivalents for biosynthesis or respiration. Many chemolithotrophic pathways are now known in which a wide range of inorganic substances, such as H_2S, NH_3 (ammonia), Fe^{2+} (ferrous iron), etc., are used by bacteria as a source of energy. For a range of reviews on this subject, see [23, 24, 28, 39, 42, 85]. Winogradsky then moved to the Swiss Polytechnic Institute in Zurich in 1888 where he developed his physical chemistry and bacterial skills in association with Ernst Schultz and Otto Roth (Winogradsky had a very poor opinion of Roth, see the quote in [17]). At Zurich, he studied the chemolithotrophy of sulfur and iron bacterium. Winogradsky also began his '*monumental studies on nitrification that provided clear evidence for the process of bacterial autotrophy and for its roles in the cycles of Nature*' [17] for which he is most famously remembered. He was the first to isolate and characterize nitrifying bacteria that oxidized NO_2^- (nitrite) to NO_3^- (nitrate), and confirmed that the oxidation of nitrogen in soil was a two-step biological process: NH_4^+ (ammonium) to NO_3^- and NO_3^- to NO_2^- (nitrite), with each step being performed by different organisms. This finding led directly to the concept of sulfur and nitrogen cycling in nature. For recent reviews on nutrient cycling and the roles of microorganisms, see [11, 20, 22, 24, 47, 49, 76, 85].

Winogradsky also reported the first example of chemoautrophy in which the energy gained from NO_2^- oxidation is used to fix CO_2 (carbon dioxide) to make organic compounds for growth or energy. The significance of this was that it was the first demonstration as an alternative to photosynthesis in which the energy from light is used to fix CO_2 (for a description of the different modes of energy conversion used by bacteria, see [79]). Dworkin [17] suggests that at this stage in the 1890's, Winogradsky began to recognize the need to re-examine and reinterpret laboratory observations with what actually happens in soils and elsewhere, marking the true beginning of the field of microbial ecology. It is noteworthy that Ernst Haeckel first defined ecology (œkologie) in 1868 as '*the science of all the relations of organisms with the external world, with the organic and anorganic conditions of existence; ... the mutual relations of all organisms living in one and the same place, their adaptation to environmental circumstances, their transformations in the struggle for life ...*' (Haeckel, quoted by [4]). Winogradsky returned to St. Petersburg in 1891 to be the Head of the General Microbiology Department of the Imperial State Institute of Experimental Medicine,

[8] He also determined that S^0 was further oxidized to SO_4^{2-} (sulfate) to release even more energy. The SO_4^{2-} is excreted as H_2SO_4 (sulfuric acid).

eventually becoming the Director in 1902. During this period he made a number of important discoveries, including the first isolation of the free-living anaerobic nitrogen fixer, *Clostridium pasteurianum*, originally known as *C. pastorianum*. This anaerobic bacterium could grow under N_2 (nitrogen) gas but not under normal aerobic (oxygenated) conditions. He subsequently showed that his original isolation of *C. pasteurianum* in flasks was successful because he had purified it in the presence of two contaminating aerobic organisms. These were able to reduce O_2 levels sufficiently to allow *C. pasteurianum* to grow anaerobically. In soils, the co-existence of different species having different nutritional requirements and activities underpins community function, and once again, Winogradsky's experiments with simple microcosms demonstrates that important ecological interactions can be identified and experimentally tested. In 1896 he presented a lecture at the Institute of Experimental Medicine '*On the role of microbes in the general cycle of life*' in which he visualized the activity of microbes as regulating the circulation of chemicals between the organic and inorganic realms throughout the world [2].

In 1905 Winogradsky resigned as Director of the Imperial State Institute of Experimental Medicine and retired to his estate in Gorodok in the Ukraine. However, he was forced to flee in the aftermath of the Russian Revolution in 1921. He spent the last thirty years of his life at the Pasteur Institute, which at that time was the centre of European microbiology, '*formulating the ideas and precepts of what was to become known as microbial ecology*' [17]. Central to this was the 'Direct' method which included microscopy and microphotography to '*gain direct access to the microscopic landscapes of soils*' [2]. In Figure 1 we present a soil landscape drawn by Winogradsky published in the 1920's[9] showing the distribution of microorganisms and particles of soil in a sample obtained from the Brie region of France [105, 106] (his soil landscapes have an artistic appeal that are lacking modern soil images, *cf.* Figures 5, 6 and 7, but in turn, his images lack the quantitative detail expected in contemporary research). During this period, another early pioneer of soil microbiology working in Kyiv, Nikolai Cholodny, developed the famous buried glass slide method to sample soil microorganisms whilst maintaining some spatial detail [12, 13]. He also realized that soil microorganisms could be studied by microscopy under natural conditions, using small transparent soil chambers where the influence of soil treatments could be studied experimentally. The innovative aspect of Winogradsky's Direct method, including contributions from others such as Cholodny, was that it did not rely on *in vitro* cultures but rather allowed investigation of microorganisms directly in the complexity of their natural environment. Edward Russell working at the Rothamsted Agricultural Experimental Station in the UK[10] corresponded with Winogradsky whom he called '*the founder of soil microbiology*', discussing the application of the Direct method in their work. Russell's colleague, Henry Thornton, began to combine it with their statistical approach to investigating soil microbiology. However, to link microbial distributions in soils with biochemical processes, Thornton maintained that there was still an essential place for the investigation of *in vitro* cultures in soil microbiology [2], a view that was ultimately accepted by Winogradsky in the development of his Direct method. During his career, Winogradsky was transformed from plant physiologist, to microbiologist and then to the first ecological

[9] It is worth mentioning that the first 35mm camera produced by Leica was not available until 1925, so hand-drawn microscopic observations were routinely used at the time of Winogradsky's soil studies.

[10] Apropos nothing at all, Andrew Spiers has a link with Rothamsted as some relatives live nearby.

microbiologist [3, 17]. In 1949 he published *Microbiologie du Sol: Problèmes et Méthodes, Cinquante ans de Recherches* (Soil Microbiology: Problems and Methods, Fifty Years of Investigations) [105, 106], a compendium of his research which included an essay describing the synthesis of microbiology and ecology into the new discipline of microbial ecology, and the development of soil microbiology. In this, his direct method had evolved into a synthetic approach that accommodated the biological, chemical and physical characteristics of soils. He argued that to study soil microbiology, a combination of techniques were required and that the biological, physical and chemical components should be thought of as parts of a single system [2]. In his biography Waksman concludes, in a somewhat understated manner, that Winogradsky had made '*significant contributions to the sciences of plant physiology, microbiology, soil science, and ecology*' and that '*his work unites these fields in a novel way*' [104]. Winogradsky published his last paper in 1953 and died the following year aged 97.

From Geissler Chambers to the Winogradsky Column

Winogradsky began his observations of *Candida vini* using Geissler chambers which allowed him to make prolonged microscopic observations of the growing cells for up to eight months at a time. The Geissler chamber[11] was a section of glass tube that had been flattened to provide a small growth zone or microcosm through which different media could be passed (see [2, 17, 61]). These were probably the first flow chambers for the direct observation of microorganisms, very similar to the pedoscops capillary system proposed almost a century later [65], and analogous to current flow systems used in combination with CLSM (Confocal laser scanning microscopy) (e.g. [95]). Winogradsky went on to use larger microcosms to study sulfur bacteria, inoculating 3 – 5 litre flasks with pieces of muddy *Butomus umbellatus* (flowering or grass rush) rhizome which were then maintained for up to 6 weeks whilst measuring H_2S production and sampling for bacteria. He also used small glass cylinders and Petri dishes containing silica gels soaked with different nutrients to maintain and examine soil cultures. Winogradsky used glass cylinders filled with sulfur spring water and sediments to grow *Beggiatoa.* He subsequently used them to describe a variety of organisms, including purple photosynthetic bacteria[12]. These cylinders, now known as Winogradsky's columns, containing various mixtures of sand, soil, sediments, and water, are ideally suited to introducing microbiology students to selective culture methods, microbial community succession, and stratification along chemical and physical gradients (e.g. [6, 74, 77, 78]).

Winogradsky's columns are easily constructed and microbial changes are observable within weeks (as an example, see Figures 2 and 3 showing the construction of a pair of columns using river water and sediments collected from the banks of the Dnipro in Kyiv and their development over a period of several weeks). A further refinement of the Winogradsky column includes the use of agar to semi-solidify aqueous layers containing different nutrients to allow the isolation of bacteria.

[11] The 'Geissler tube' was the name given to early gas discharge tubes invented by the physicist Heinrich Geissler in 1857. It is not clear how the Geissler chamber used by microbiologists might relate to this, though it was used by Pasteur in the 1860's [61].

[12] Winogradsky was unaware at the time that they were capable of anaerobic anoxygenic photosynthesis [78].

Figure 2. Constructing Winogradsky's columns. We produced Winogradsky's columns using river water and sediments collected from the banks of the Dnipro. (Top left) The snow and ice-covered Dnipro in Kyiv in January 2013. (Bottom left) Sediment samples diligently collected by two of the Authors. (Right) Mariia Boretska (left) and Olena Moshynets (right) adding sediments and water to a glass cylinder to produce the Winogradsky's column. See Figure 3 for a series of images taken of these columns over a period of several weeks. Images are from Mariia Boretska and Olena Moshynets.

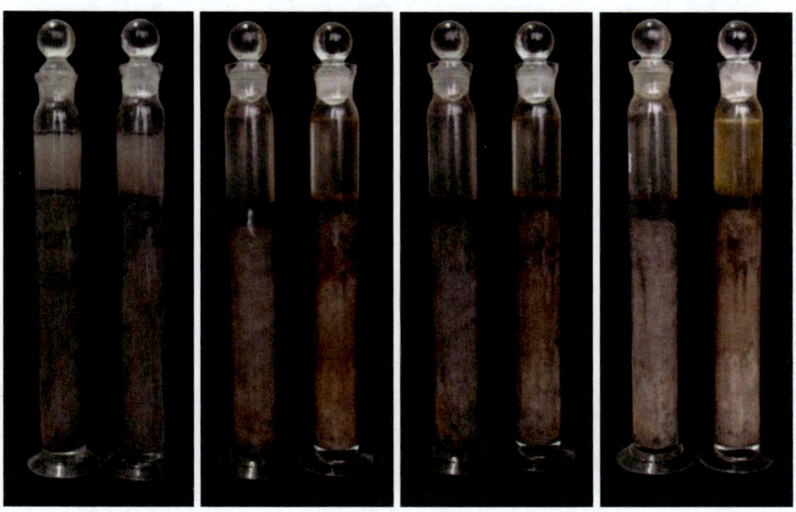

Figure 3. Microbial development of the Winogradsky's columns. The Dnipro river sediment and water columns were maintained at room temperature (~20°C) for several weeks under anaerobic conditions during which microbial colonisation and stratification occur. Shown here are a series of photographs of two columns; the left-hand one was incubated in the dark, and the right-hand one exposed to natural light. (Left to right) Immediately after construction; after 13, 23 and 37 days. Images are from Mariia Boretska and Olena Moshynets.

In a study of the cycling of H_2S and SO_4^{2-} in Black Sea coastal waters, another early pioneer of environmental microbiology, Mikhail Egunov, observed the development of a bacterial film[13] between the upper SO_4^{2-}–rich layer and the lower H_2S–rich layer[14], and concluded that such a region occurred in the Black Sea at a depth of 200 m [18]; a sketch of Egunov's column was reproduced in a later publication by Winogradsky's only student, Vasily Omeliansky [57]). Although Winogradsky obtained his *Beggiatoa* from sulfur springs, they have also been recovered from the Black Sea and other marine environments, including deep-sea hydrothermal vent environments where they oxidize H_2S to S^0 and S^0 to SO_4^{2-}. In a study of the mechanism and rate of sulfide oxidation in water, Yury Sorokin used an experimental column to model the vertical gradient of the oxidation-reduction (redox) potential of sea water and bacteria in the Black Sea in which he also cited Egunov's earlier work [87]. After a period of two weeks Sorokin observed the stratification of bacteria in two principal regions of the column. The upper one contained H_2S–oxidizing *Acidithiobacilli*, formerly known as *Thiobacilli*, between the upper O_2–rich water layer and the lower H_2S–rich anaerobic water layer. The lower region was directly above a layer of agar at the bottom of the column, containing sulfide and sulfates which diffused upwards into the water column. In this lower region, sulfate-reduction was observed by the development of a ring of black sulfate-reducing colonies on the cylinder walls. Sorokin's column supported all of the biological functions required for a complete sulfur cycle, despite being structurally a very simple microcosm. Investigations of Black Sea microbiology continues today, almost 120 years after Egunov's original work, and now includes modern approaches such as metagenomics (reviewed by [96]). The ability to obtain bacterial sequences from environmental samples, such as particulate matter or marine snow, can provide information on bacterial diversity and community function. Sequences obtained from snow samples recovered from the suboxic zone of the Black Sea suggest that the interior of these sinking aggregates might support Mn (manganese) and SO_4^{2-} reduction as well as H_2S and S^0 oxidation [25]. Although the redox gradient might be the most obvious large-scale chemical gradient in the Black Sea, it is clear that there are smaller, local chemical gradients that might also affect bacterial activity and help define micro-habitats. Micrometre–scale gradients also exist within sediment and soil pore networks and aggregates where they help define micro-habitats and community function.

Contemporary Use of Microcosms in Bacterial Research

It is humbling to summarize the achievements of the founder of contemporary ecological microbiology, albeit in such a superficial manner and without direct reference to his scientific publications, but it is also a useful exercise to show the early development and use of microcosms in microbiology. A microcosm can be simply defined as '*a community, place, or situation regarded as encapsulating in miniature the characteristics of something much larger*' [63]. They have been used by biologists from a variety of disciplines to study physiology, processes, and interactions over a range of scales, from single cell-traps to laboratory cultures, aquaria, field trials, to the management of national parks, entire river systems, lakes, coastal regions, and islands.

---

[13] A 'bacterial plate' may be a more accurate translation of 'бактериальная пластинка' [18].

[14] This is similar to Winogradsky's observations of *Beggiatoa* behaviour under a cover-slip.

Figure 4. Microcosms have been used to investigate aspects of microbiological ecology, adaptation and evolution. (Top left) Plant-specific induction of *Pf.* SBW25 gene expression has been examined using sugar beet seedlings grown in a vermiculite – compost mixture (the plastic tube has a 1 cm diameter). (Top middle) Bacterial evolution and biofilm-formation have been investigated using glass vials containing liquid growth media (the glass vial has a 2.75 cm diameter). The microcosm on the left contains wild-type *Pf.* SBW25 that grows in the liquid column, and in the microcosm on the right, the evolved Wrinkly Spreader mutant forms a biofilm at the air-liquid interface. (Top right) The impact of bacterial surfactant expression on water distribution within soil pore networks has been investigated using sterile repacked soil aggregate cores (the core ring has a 37 mm diameter). (Bottom) Soil pore structures visualized by X-ray CT have been printed out at a larger scale in plastic to allow the investigation of fungal and bacterial colonization processes (the smallest cube is 1.8 cm wide). Images are from Andrew Spiers and Ruth Falconer.

Although the use of microcosms in research has important caveats, most often associated with the limited complexity of the system or the application of results to large-scale systems, they nonetheless provide a very powerful experimental system for biological research. As an example of this, we take the opportunity here to briefly present the research that colleagues and ourselves have undertaken in this field over the last 15 years, utilizing artificial growth media in glass vials and petri dishes, plants and repacked soil aggregate cores, and most recently, plastic and transparent soils, to investigate aspects of microbial ecology, adaptation and evolution (see Table 1 and Figure 4). Much of this work has involved the use of the soil and plant-associated fluorescent pseudomonad, *Pseudomonas fluorescens* SBW25 (referred to henceforth as *Pf.* SBW25), as a model organism in experimental evolution and biofilm studies (e.g. [44, 72, 92, 93]) and plant-microbe interactions (e.g. [26, 27, 69, 83]). *Pf.* SBW25 was isolated from a leaf of a sugar beet crop grown at the University Farm at Wytham, Oxford, UK (SBW stands for <u>S</u>ugar <u>B</u>eet <u>W</u>ytham) [71]. Like other fluorescent pseudomonads, *Pf.* SBW25 is a benign, plant growth-promoting bacterium, colonizing the leaves and roots of a wide variety of crop plants and weeds. In general, *P. fluorescens* strains are not considered plant pathogens [64], though like some other plant growth-promoting bacteria, *Pf.* SBW25 carries a defective pathogen-like type III secretion system [66].

Table 1. Contemporary investigations

Investigation	Microcosm	Sampling and measurement	Potential variables
Experimental evolution[a] and biofilms[b.]	Glass vials containing liquid growth media.	Destructive sampling to determine diversity, fitness, biofilm growth, strength and attachment.	Nutrients, O2, physical disturbance, liquid viscosity, interspecific completion, phage and plasmid co-evolution.
Plant-specific induction of bacterial gene expression[c].	Sugar beet seedlings grown in compost-vermiculite or potting mixture.	Destructive recovery of bacteria and sequencing; re-testing to determine expression levels and fitness; Raman spectroscopy.	Plant, bacteria and soil type, temperature, water and humidity.
Fitness advantage of cellulose expression.[d]	Surface-sterilized mushroom caps.	Destructive recovery of bacteria to determine fitness.	Mushroom and bacteria type, presence of predators and competitors.
Impact of bacterial surfactants on soil hydrology.[e]	Repacked soil aggregate cores.	Destructive determination of water content.	Bacteria type and distribution, soil type, porosity, water content.
Fungal invasion of soil pore networks.[f]	Printed plastic model of real soil structure.	Non-destructive photography and X-ray CT imaging.	Structure (porosity, pore dimensions, connectivity, etc.).
Transparent soil structures.[g]	Growth of plant roots in Nafion beads.	Non-destructive optical and X-ray CT imaging; root trajectories and contact points, bacterial distribution and activity.	Plant and bacteria type, structure (porosity, pore dimensions, connectivity, etc.), Nafion surface chemistry, pore liquid chemistry and nutrients.

Investigations led by: (a), Angus Buckling, Michael Brockhurst, Rees Kassen and Paul Rainey; (b), Paul Rainey and Andrew Spiers; (c), (d) and (e), Andrew Spiers; (f) and (g), Ruth Falconer and Wilfred Otten.

Furthermore, many *P. fluorescens* strains including *Pf.* SBW25[15] produce soft rot-like symptoms in plant tissues following physical damage, suggesting opportunistic pathogen-like tendencies and a possible pathogenic origin for the species. Furthermore, the ability to colonize new environments is a characteristic associated with the pseudomonads, and may be enabled by relatively large genomes that include many sensory-regulatory elements [84, 90]. The *Pf.* SBW25 genome has been fully sequenced [83], and it can be manipulated easily using standard molecular biology techniques.

Finally, the close similarity with the opportunistic human pathogen *P. aeruginosa*, the plant pathogen *P. syringae*, and the archetypal soil isolate *P. putida*, make *Pf.* SBW25 an ideal model organism for experimental studies.

[15] Andrew Spiers, unpublished observations.

Plant Microcosms

The ability of *Pf.* SBW25 to successfully colonise leaf and root surfaces depends on the close adaptation of the bacteria to the local micro-scale chemical and physical conditions prevailing on the plant, referred to as the phyllosphere, and in the narrow layer of soil in close association with the roots, referred to as the rhizosphere. The ecological performance and fitness of *Pf.* SBW25 in these different and dynamic environments is a complex phenotype, in part explained by the ability of the bacteria to adjust gene expression patterns, enzyme activities, physiology, metabolism, and behaviour in response to environmental opportunities and signals. Various techniques have been developed to examine plant-specific induction of bacterial gene expression, and one of these, known as IVET (*in vivo* expression technology), was used by Paul Rainey's group[16] to investigate *Pf.* SBW25 gene expression on sugar beet seedlings in simple microcosms (see Figure 4) [26, 27, 69, 83]. In this work, the microcosm consisted of small plastic tubes containing vermiculite or vermiculite-compost mixtures in which individual sugar beet seeds were germinated and grown for 2 – 4 weeks. This period of growth provided a selective environment in which only those IVET constructs responding to specific, but unknown, plant signals survived.

The key to the IVET strategy was the use of *Pf.* SBW25 mutant strains that had been deleted for an essential gene, either *dapB* needed to produce diaminopimelate required for the cell wall, or *panB* required to produce the vitamin pantothenate, and a library of IVET plasmids each containing a random fragment of the *Pf.* SBW25 genome inserted immediately up-stream of a second copy of the deleted gene [26, 69]. If the genome fragment contained a promoter sequence in the correct orientation, the plasmid-born copy of the essential gene might be expressed allowing the host bacterium to survive[17]. Libraries of bacteria were used to inoculate sugar beet seeds, and bacteria were recovered from the leaves, roots and bulk soil after 2 – 4 weeks. These bacteria were tested on agar plates to exclude those plasmids containing constitutively-active promoter sequences which had no environmental relevance. The genome fragment in each IVET plasmid responding to specific environmental signals was mapped back to the *Pf.* SBW25 genome by sequence analysis to determine which genes were being expressed. Further experiments were conducted to assess the levels of gene expression and their likely function in the sugar beet environment. Examples of genes identified in this manner are those involved in nutrient acquisition, stress responses, and the expression of antibiotics, cellulose and phytohormones [26, 27, 69, 83].

Unrelated research by the Rainey group investigating the *Pf.* SBW25 mutant Wrinkly Spreader (WS) biofilm had just resulted in the identification of the cellulose biosynthesis operon (*wss*) [92] and a demonstration that *Pf.* SBW25 could expressed partially-acetylated cellulose in an air-liquid interface biofilm [92, 93] (discussed further in the following section; see also the review by Spiers et al. [91]). Both biofilm-formation and the involvement of cellulose in these structures is common amongst environmental *Pseudomonas* spp., giving rise to the suggestion that the function or ecological role of biofilms in natural environments might be to localize bacteria at the air-liquid interface in small bodies of water or liquid such as those found in soil pores, on plant surfaces, or in plant tissues [73, 89, 91, 101]. Fitness

[16] Department of Plant Sciences, University of Oxford, UK.

[17] The genome fragment might contain i) no promoter activity; ii) a promoter only active in the plant or soil environment; or iii) a promoter constitutively active in all environments including agar plates.

assays had shown that there was an advantage to cellulose-expression by wild-type *Pf.* SBW25 on sugar beet leaves and roots, but not in bulk soil [26], though this type of assay does not provide a mechanistic explanation of how cellulose might improve fitness in these environments. Recently, however, investigation of cellulose-expression in the related pseudomonad, *P. putida* mt-2, suggest that it may provide a fitness advantage in water-limiting conditions [53]. Examination of the survival and fitness advantage of a set of Brown blotch-causing pseudomonads isolated from mushrooms, plus *Pf.* SBW25 and *P. putida* KT2440 (a derivative of *Pp.* mt-2), support the view that cellulose might provide an advantage by retaining water in micro-colonies and the absorption of water directly from vapour under conditions of water stress, rather than by the formation of large biofilm structures [44, 91]. The response of *Pf.* SBW25 to plant surfaces has also been examined by single-cell Raman micro-spectroscopy [31]. This technique provides spectra reflecting the whole-cell chemical composition of bacterial cells, and provides a means of monitoring growth and metabolic status. Raman analysis is normally applied in the identification of bacteria (e.g. [46]), but it might be used in a different manner to determine changes in metabolism and physiology for bacteria growing on different nutrients or environments [33]). *Pf.* SBW25 cells grown *in vitro* in defined minimal media containing single nutrient sources, such as simple sugars, amino acids and TCA (tricarboxylic acid) cycle intermediates, were found to show significantly different Raman spectral 'fingerprints', suggesting that the technique could be used to differentiate bacterial cells growing in natural conditions on the basis of local micro-scale chemical and physical conditions, growth history and strategy. For example, this type of Raman analysis could differentiate between cells growing on fructose or glucose (or glycerol), arginine or aspartic acid (or asparagine), etc., and could be used to follow cells starving over a period of nine days [31], cells growing in high or low–O_2 conditions or in a biofilm, and cells carrying the environmental *Pseudomonas*-associated plasmid pQBR103 [36, 102].

Raman analysis could also differentiate between *Pf.* SBW25 cells growing in the sugar beet rhizosphere and bulk soil without plants, suggesting that *Pf.* SBW25 responds to different environmental signals which result in significant shifts in physiology and metabolism.

Previously, it had been suggested that a correspondence between Raman spectra obtained for bacteria growing in a natural environment and the same strain on defined growth media *in vitro* might be used to infer the metabolic status of the former [33]. *Pf.* SBW25 recovered from sugar beet leaves were found to have Raman spectral profiles more similar to *Pf.* SBW25 grown *in vitro* on fructose or aspartic acid, rather than on glucose or arginine, and where quite dissimilar to starved cells [31]. This suggests that *Pf.* SBW25 growth on the leaves of sugar beet is likely to be neither carbon-catabolite-repressed nor carbon-limited. Wei Huang et al.[18] have also demonstrated the utility of Raman analysis further by showing that it can be combined with both SIP (stable isotope probing) and FISH (fluorescence *in situ* hybridization) [32, 35]. Raman can also be combined with laser optical tweezers to trap and recover individual bacteria from solution [34, 107].

---

[18] Centre for Ecology and Hydrology, Oxford, UK.

Experimental Evolution and Biofilms

Experimental evolution has been used by ecologists in an attempt to understand the dynamics of diversifying and adapting populations, and more recently, in an attempt to link evolutionary dynamics, fitness, and niche specialisation, with an understanding of the mechanistic link between the underlying organismal molecular biology and environmental factors (for reviews, see [10, 41, 70]). *Pf.* SBW25 has been used as a model organism in experimental evolution studies using simple glass microcosms and was first described by Paul Rainey and Michael Travisano [72] (see Figure 4; see also Chapter 5 and references therein). When wild-type *Pf.* SBW25 is inoculated into nutrient-rich King's B liquid medium [43] the population will grow over a period of several days and accumulate mutants. One of the most visually obvious of these mutants is the Wrinkly Spreader, characterized by a wrinkled colony morphology on King's B agar plates. The Wrinkly Spreader also differs from the ancestral, wild-type *Pf.* SBW25 in that it no longer colonizes the liquid column of static King's B microcosms, but grows at the air-liquid interface by forming a robust biofilm (for a description of the evolution of the Wrinkly Spreader, see [29]). Furthermore, in static microcosms, the Wrinkly Spreader has a fitness advantage over the ancestral *Pf.* SBW25 and other non-biofilm–forming competitors [29, 92]. This advantage demonstrates that the Wrinkly Spreader is not simply one of many different random mutants that might arise during the diversification of *Pf.* SBW25 populations, but is an evolved genotype, representing a fitness or adaptive leap from the ancestral *Pf.* SBW25 and having a new niche preference for the air-liquid interface formerly not colonized by *Pf.* SBW25. In shaken microcosms where biofilms cannot form, the Wrinkly Spreader does not have a fitness advantage over wild-type *Pf.* SBW25 or other non-biofilm–forming strains [29, 92], and on agar plates, the Wrinkly Spreader is at a disadvantage and is genetically unstable [88]. Investigation of the distribution of O_2 in static microcosms by Andrew Spiers' group[19] has shown that the *Pf.* SBW25 colonisers modify the environment by creating an O_2 gradient within hours of introduction into static King's B microcosms [45]. The O_2 gradient divides the microcosm into an upper, O_2-rich zone of ~200 μm, and a lower O_2-depleted zone. These zones present the developing, diversifying *Pf.* SBW25 population with new niches to occupy. Although Wrinkly Spreaders may appear anywhere in the liquid column, they are more likely to appear in the O_2-rich zone as this supports faster growth; once they appear and localize to the air-liquid interface, they begin to produce a biofilm which will intercept O_2 diffusing into the liquid column from the air above. By intercepting this O_2, Wrinkly Spreader cells at the air-liquid interface will grow faster than the ancestral wild-type *Pf.* SBW25 and other non-biofilm–forming mutants lower down in the liquid column, thus explaining the fitness advantage of the Wrinkly Spreader [45]. As the Wrinkly Spreader biofilm develops, there is opportunity for further evolution as the O_2 gradient now subdivides the biofilm itself into an O_2-rich surface layer of ~100 μm and a lower O_2-depleted zone.

In a similar air-liquid interface biofilm produced by *Gluconacetobacter xylinus* (formerly known as *Acetobacter xylinum*), the opposing O_2 and nutrient gradients combine to define an optimal growth zone 50 – 100 μm deep, where there is sufficient O_2 diffusing down from the air above and nutrients diffusing upwards from the liquid column [103]. Recent experiments by the Spiers' group suggest that the fitness advantage of the Wrinkly Spreader in static

[19] The SIMBIOS Centre, University of Abertay Dundee, UK.

King's B microcosms also depends on the relative availability of O_2 and nutrients, as well as on the relative expense of biofilm-formation versus the cost of maintaining position at the air-liquid interface by swimming motility. Growth in O_2-rich conditions, and in a biofilm rather than as individual free-swimming planktonic cells, has a big impact on the metabolic and physiological status of bacteria, as shown by single cell Raman analysis of *Pf.* SBW25 [36] and a similar analysis of a different *P. fluorescens* strain investigated by ATR–FTIR (attenuated total reflection – Fourier transform infrared spectroscopy) [67]. The relative ease in distinguishing Wrinkly Spreader colonies from wild-type *Pf.* SBW25 and other mutant classes on King's B agar plates has meant that studying diversification and adaptation is relatively simple[20]. For example, the impact on physical disturbance, nutritional complexity, and the presence of bacteriophage[21] have all been examined using this simple microcosm system [8, 9, 40, 62]. The molecular biology underlying the WS (Wrinkly Spreader) phenotype, including the colony morphology and biofilm, has also been examined in detail [7, 92, 93, 94]. Using a mini-transposon mutagenesis approach, a number of mini-Tn*5* mutants of the archetypal Wrinkly Spreader producing wild-type *Pf.* SBW25–like colonies and unable to form biofilms, were isolated and the position of the mini-transposon located on the genome determined by sequencing. This approach and subsequent testing determined that the WS phenotype required the expression of partially-acetylated cellulose and an unidentified pili-like attachment factor [92, 93, 94], and that it was activated by a single amino residue change in a regulatory sub-unit of a membrane-associated chemoreceptor [7]. This mutation activated a diguanylate cyclase known as WspR, to produce cyclic-*di*-GMP (cyclic *di*-3',5'-guanylate) which then activated the cellulose synthase complex and the expression of the attachment factor [7, 48] (for a more detailed description of the WS phenotype see [29]; for a review of cellulose-expression and biofilm-formation amongst the pseudomonads, see [91]; and for a review of cyclic-*di*-GMP signalling in bacteria, see [38]).

SIMBIOS and Soil Research

The Scottish Informatics, Mathematics, Biology and Statistics (SIMBIOS) Centre was established at the University of Abertay Dundee to support a range of interdisciplinary research activities, including investigations of the micro-scale structure of soil and the use of modelling approaches, to understand how spatial heterogeneity at the microscopic scale affects biophysical behaviour at larger scales. Key to this work has been the use of high resolution X-ray CT (micro-X-ray computed tomography) to non-destructively visualize and quantify soil pore networks, resolving features down to 3 μm resolution and differentiating between materials of the basis of electron density (reviewed by [97]). A 2-D (two-dimensional) image of the physical structure of a soil aggregate, and 3-D (three-dimensional) images of a soil sample visualized by X-ray CT are provided in Figure 5 (images such as these are usually manipulated on screens where they can be rotated in 3-D space allowing the viewer an excellent impression of pore spaces and networks). SIMBIOS uses X-ray CT to investigate micro-habitat structures in natural soil and repacked soil aggregates, and are

[20] Experimental evolution is studied using other bacteria which do not produce Wrinkly Spreader-like mutants. In these systems, more laborious or expensive procedures must be used to distinguish between mutants.

[21] A bacterial virus.

developing in parallel new statistical and image analysis tools required for this type of work [54, 56]. They are also combining the 3-D microtomographs of soil structure with other techniques, such as aggregate sampling and thin sectioning used to investigate bacterial colonization (e.g. [55, 56]) and SEM–EDX (scanning electron microscopy – energy-dispersive X-ray spectroscopy) to map elemental distributions [30, 58]. Natural soil structures are hard if not impossible to replicate for research purposes (e.g. example, to investigate aspects of bacterial colonization, community structure and function, plant root development, and plant-bacterial interactions).

Soil microcosms are generally constructed by re-packing sieved soil aggregates (see Figure 4), and X-ray CT can be used to quantify pore structure in terms of porosity, pore dimensions and connectivity, etc., of these artificial environments. However, X-ray CT is limited by current ability to differentiate between materials of similar electron density (e.g. differentiating the air, water, organic and inorganic matter can be problematic). Under some circumstances, X-ray CT may be able to visualize water films in soil pore networks (e.g. [99]), and in the near future, it may be possible to map the spatial distribution of water within the soil pore network at the micrometre scale.

Figure 5. X-ray CT can be used to visualize the physical structure of soils. Grey-scale images are used to indicate different densities ranging from air-filled pore spaces (white) to dense aggregates (grey) and solid particles (black). (Left) A 2-D slice showing the internal structure of a soil aggregate (2 x 2 mm) at a spatial resolution of 3 μm. (Right) A 3-D projection of a soil sample (1.4 x 1.4 x 2.7 cm) is shown as a grey-scale solid and again showing only the pore spaces, at a spatial resolution of 32 μm. Images are from Wilfred Otten.

Plant roots in soil can also be visualised by X-ray CT [100], and Sonja Schmidt et al.[22] have used it to investigate the interaction between roots and soil particles at the micrometre scale, finding that the level of root-soil contact increased with decreasing particle size, and possibly also as a result of changes in particle packing around the root [80].

[22] The SIMBIOS Centre, University of Abertay Dundee, UK, and the James Hutton Institute, Invergowrie, UK.

Soil Microorganisms

The glass slides used by Cholodny [12, 13] and other early investigators to visualize soil microbiology whilst retaining some spatial detail are largely forgotten today as focus has shifted to the use of molecular techniques to investigate community function and diversity. However, there is growing recognition that there is a need to investigate microbial community function at the micrometre in the soil pore networks, where local physical and chemical conditions, and inter-species interactions are key to community function. The glass slides used by Cholodny are often viewed as problematic, insofar the slide represents a large, solid, and impermeable surface that alters soil structure and might restrict plant root and fungal development, and microbial mobility. However, these problems are reduced by using smaller sampling devices with different surface chemistries and permeabilities. Plastic films such as PET (polyethylene terephtalate) can be used instead of glass slides, and are readily cut into smaller samplers of 2 – 5 mm^2.

Olena Moshynets et al.[23] have shown that both *Pf.* SBW25 and *P. putida* KT2440 readily attach and detach from PET films *in vitro*, and pre-conditioning with soil wash suspensions can be used to modify the PET surface to increase bacterial attachment [50]. PET films buried in soil microcosms are rapidly colonized by microorganisms which could be investigated by CLSM or recovered onto agar plates; examples of this are shown in Figure 6 (like the soil landscapes produced by Winogradsky, these images are visually more exciting than the X-ray images and give a good impression of the microbiological complexity of soils at the micrometre scale). PET film samplers buried in soil microcosms can be non-destructively imaged by X-ray CT to map soil aggregate–film contact points and pore spaces [50]. In this manner, it may be possible to link specific micro-habitats identified by X-ray CT with molecular analyses conducted on the recovered film still retaining spatial detail (e.g. FISH or PCR (polymerase chain reaction) approaches to identify species and to examine gene expression patterns). However, some techniques such as Raman may not be readily applicable due to the auto-fluorescence spectrum of the PET polymer[24]. The PET film samplers have also been used to investigate bacterial colonization of bamboo tissues and the rhizosphere of *Brassica napus* (rapeseed) [51, 52]. Archana Juyal and Thilo Eickhorst[25] are investigating the dynamics of bacterial colonization of soil pore networks. They are using repacked soil aggregate microcosms inoculated with model microorganisms such as *Bacillus subtilis* and *Pf.* SBW25. These microcosms are embedded in resin and thin sections prepared for microscopy.

Two examples of the images obtained using this technique are provided in Figure 7; in the first, all microorganisms are stained using a DNA-binding fluorescent dye, and in the second, *Betaproteobacteria* cells are specifically visualized by FISH (both images are of the same region of the thin section). Although the embedding process can result in sample shrinkage, they are able to retrospectively identify the position of sliced sections in X-ray CT images of the original core, similar to the mapping of buried PET film samplers. Thilo Eickhorst is also developing CARD-FISH (catalyzed reporter deposition–FISH) for use with

---

[23] Institute of Molecular Biology and Genetics, National Academy of Sciences of Ukraine, Kyiv, Ukraine, and during visits to the SIMBIOS Centre, University of Abertay Dundee, UK.

[24] Wei Huang, unpublished observations, personal communication to Andrew Spiers.

[25] The SIMBIOS Centre, University of Abertay Dundee, UK, and the Institute of Soil Science, University of Bremen, Germany, respectively.

the thin sections, as this technique promises to be a more sensitive method than standard FISH or the use of fluorescent stains that bind to DNA (deoxyribonucleic acid) to detect the presence of microorganisms [5, 19, 81].

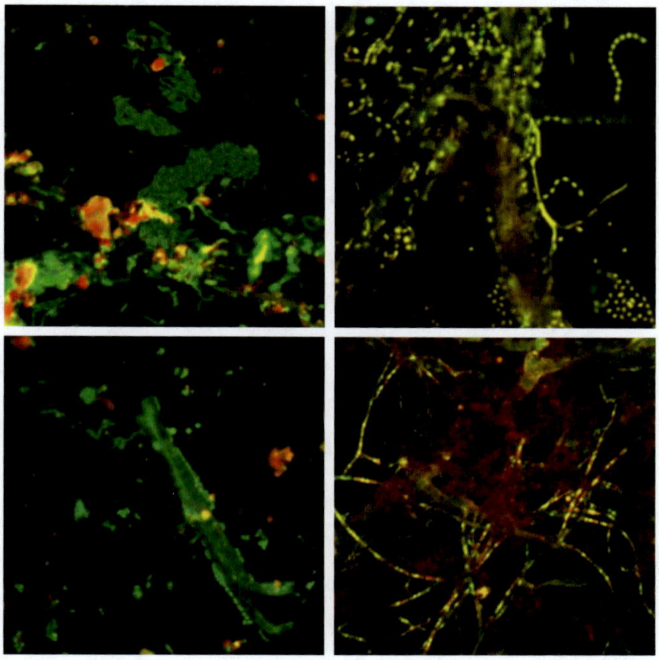

Figure 6. PET film samplers can be used to visualize the microbiological complexity of soils. Films were buried in soil for two weeks before recovery and staining with the DNA-binding fluorescent dye, Acridine orange (3,6-acridinediamine), to visualize microorganisms. Sections of film surfaces (60 x 60 μm) were inspected by CLSM (Confocal laser scanning microscopy), and in these images, many different types of microorganism can be distinguished but not identified. Images from Olena Moshynets.

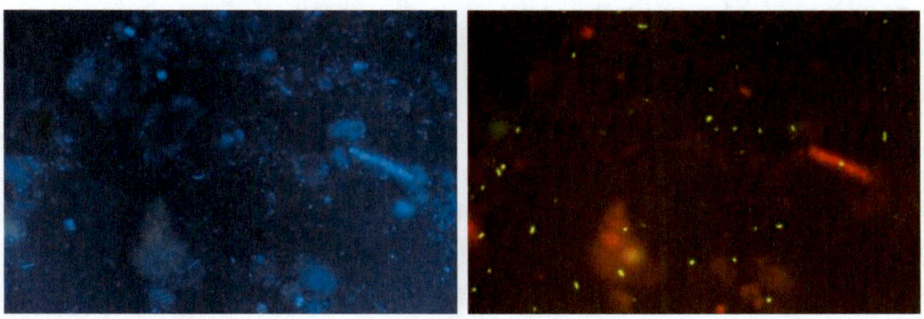

Figure 7. Bacteria can be identified in pore spaces in thin sections of soil embedded in resin. Fluorescent microscopy was used to image a region of a thin section of resin-embedded paddy soil (sandy loam) using two different techniques for the same field of view (100 x 150 μm). (Left) All microorganisms are visualized using the DNA-binding fluorescent dye, DAPI (4',6-diamidino-2-phenylindole). (Right) *Betaproteobacteria* are visualized by FISH using a specific 16S rRNA probe. Images from Thilo Eickhorst.

Although generally 16S rRNA–targeted oligonucleotide probes are used for FISH and CARD-FISH to identify microorganisms at a number of different phylogenetic levels, it is possible to use gene-specific probes that could also indicate whether individual bacteria are expressing genes involved in biological processes at the micrometre scale within soil pores. For example, the spatial clustering of bacterial nitrogen fixation genes within soil is probably due to the separation of populations across aggregates of different size classes, caused by differences in the temporal stability of aggregates as niches for microbial communities [37]. Although the identification of thin sections in the 3-D X-ray CT images of the original core were done manually, more rigorous statistical methods have been developed by Simona Hapca et al.[26] [30].

This work originated from collaboration with Clare Wilson[27] who was studying micro-spatial heterogeneity of SOC (soil organic carbon), Fe (iron) and Al (aluminum) in soils using SEM-EDX to produce two-dimensional maps which were then combined with the 3-D X-ray CT images [58].

Impact of Bacterial Surfactants on Soil Hydrology

Pf. SBW25 and other soil and plant-associated pseudomonads are known to express powerful surfactants such as the CLP (cyclic lipopeptide) viscosin, involved in a variety of roles including motility, biofilm formation, anti-bacterial activity, the inhibition of protist and nematode predation, lysis of fungi and oomycetes, and the induction of systemic resistance in plants (reviewed by [15, 68]).

However, surfactants also have a simpler impact in water where they reduce the liquid surface tension (γ) and alter the shape of liquid bodies, especially at the meniscus or the air-liquid-solid interface. The impact of reduced γ on liquid surface geometry is significant and can be illustrated by the drop collapse assay where liquids with high–γ (e.g. water) produce uniform hemispherical drops on glass slides whilst low–γ solutions produce irregular flattened shapes.

However, the ecological value to bacteria of directly modifying or manipulating water geometry through the expression of surfactants had not been considered, despite the fact that water distribution in fully and partially-saturated pores, pore-wall surface films and droplets is a key factor in bacterial colonisation and activity in soils [14, 108]. Using repacked soil aggregate microcosms (see Figure 4), the Spiers' group demonstrated that the volumetric water content of soil was significantly lowered by *Pf.* SBW25 expressing the viscosin compared to a surfactant-deficient mutant [21]. Furthermore, six of fifteen pseudomonads isolated directly from soil shown to express surfactants *in vitro* were found to have a similar impact on water content when compared to sterile microcosms [21]. These findings indicate that surfactant-expressing pseudomonads could modify local soil-water distributions within soil pore networks, and that surfactants may therefore play a significant hydrological role in soils in addition to their recognised biological activities. As X-ray CT technology develops, it may be possible to investigate bacterial surfactant-mediated changes to local soil water distribution patterns at the micrometre scale in the future.

[26] The SIMBIOS Centre, University of Abertay Dundee, UK.
[27] School of Biological and Environmental Sciences, University of Stirling, UK.

Plastic and Transparent Soils

As an extension of the experimental investigation and modelling of fungal invasion of soil pore networks [60], Ruth Falconer and Wilfred Otten[28] have employed 3-D printing techniques to produce realistic plastic soil structures using data derived from X-ray CT of native soil cores [59] (see Figure 4).

This printing technology can produce multiple replicates of the same native structure, and by transforming the data set, related structures differing in a key parameter such as total porosity can be produced for testing.

Furthermore, the plastic provides a physically and chemically homogeneous surface, allowing the effect of soil structure to be investigated without the complication of chemical heterogeneity.

However, just like the PET film samplers, the chemical surface of the plastic soils might be modified to allow subsequent investigations of physical and chemical heterogeneity in microcosms in which structure remains constant.

In some circumstances, realistic soil structure may not be a priority where research is focusing on plant root development or bacterial-root interactions. Helen Downie et al.[29] have developed OPT (optical projection tomography) to image roots and bacteria in a novel 'transparent soil' microcosm [16]. These were constructed using Nafion particles (a sulfonated tetrafluoroethylene–based fluoropolymer-copolymer) to provide a soil-like pore network that could be partially saturated with a nutrient solution to allow the growth of *Latuca sativa* (lettuce) and *Nicotiana benthamiana* (a close relative of tobacco frequently used by plant scientists) seedlings.

Immediately before visualization, the pores were saturated with a RI (refractive index)-matched solution, rendering the particles transparent and allowing the direct imaging of roots and bacteria.

The pore structure itself can be visualized using fluorescent dyes that adsorb onto the particles, while other dyes or fluorescent protein-labelling can be used to label roots or bacteria [16].

FINAL COMMENT

Winogradsky was the father of microbial ecology, developing it as a separate discipline, distinct from and as valid as medical microbiology led by the disciples of Koch. He championed the need to study soil microorganisms in their natural habitat, and pioneered the use of microcosms in which biological processes could be investigated. Winogradsky's Direct method has matured into a truly interdisciplinary research approach that is highly topical to today's environmental microbiologists and soil biophysicists. We use this approach today, employing a variety of simple microcosm systems to examine aspects of microbial ecology, and strongly encourage other microbiologists to do likewise in their own research.

[28] The SIMBIOS Centre, University of Abertay Dundee, UK.

[29] The SIMBIOS Centre, University of Abertay Dundee, UK, and the James Hutton Institute, Invergowrie, UK.

Supplementary Table 1. The professional career of S.N. Winogradsky

1877 – 1881	Entered the Department of Natural Sciences of the Physico-Mathematical Faculty, University of St. Petersburg. He then chose plant physiology as his major, working as an undergraduate research student with the botanist-physiologist Andrey Famintsyn. Awarded a Diploma of Science in 1881 and continued his research studies with Famintsyn.
1881 – 1884	**First St Petersburg research period.** Worked with Famintsyn on bacterial fermentation. Published his first research work on the nutrition and growth physiology of the yeast *Candida vini*. Awarded the degree of Master of Science in 1884.
1885 – 1888	**Strasbourg research period.** Joined the botanist and mycologist Anton DeBary at the University of Strasbourg. He investigated the sulfur-oxidising bacterium *Beggiatoa*, proving the monomorphism of so-called 'high bacteria'. Discovered the autotrophic assimilation of CO_2 by bacteria and formulated the theory of chemolithotrophy. Developed selective (elective) conditions for bacteria cultivation.
1888 – 1891	**Zurich research period.** He developed his chemistry skills with Ernst Schultz and Otto Roth at the Swiss Polytechnic Institute in Zurich, where he investigated sulfur and iron bacterium, and bacterial nitrification. Isolated pure cultures of *Nitrosomonas* and *Nitrobacter*. Lectured at the Zurich Polytechnic Institute. He rejected an offer to join the Pasteur Institute in 1891, preferring to return to Russia.
1891 – 1905	**Second St Petersburg research period.** Became the Head of the General Microbiology Department of the Imperial State Institute of Experimental Medicine at St. Petersburg (resigned as Director in 1905 but remained a member until 1912). Corresponding Member of the Academy of Sciences of France (1902). In 1903 he and Omeliansky organized the Russian Microbiological Society, where he was a director for two years. Participated in the development of anti-plague measures in Russia. In 1894 he proved the existence of anaerobic nitrogen-fixing bacteria, isolated and described *Clostridium pasteurianum*.
1905 – 1920	Due to health problems he left St. Petersburg and returned to his estate in Gorodok, Ukraine. Honorary Member of the German Botanical Society (1907). Foreign Member of the Royal Academy of Agriculture of Sweden (1908). Corresponding Member of the Academy of Medicine of Torino (1909). Honorary Member of the Russian Microbiological Society (1910). Corresponding Member of the Microbiological Society of Holland (1911). Honorary Member of the Polytechnic Institute of Riga (1912). In 1916 he moved to Odessa. Foreign member of the Royal Society in London (1918). In 1920 he left Odessa following the Russian Revolution by French warship as a Corresponding Member of the French Academy of Sciences.
1921 – 1922	Joined the Agricultural Institute of the University of Belgrade, Yugoslavia. Member of the Czechoslovak Botanical Society and a Professor of the University of Belgrade (1921). In 1922 he accepted an invitation from Emil Roux, director of the Pasteur Institute and moved to Paris.
1923 – 1953	**Paris research period.** Continued working on nitrogen-fixation in bacteria and soils; he was the first to study soils as an environment. Developed his Direct method to investigate the biological, chemical and physical characteristics of soil, beginning the modern disciplines of ecological microbiology and soil microbiology. Head of the Laboratory of Agricultural Microbiology of the Pasteur Institute in Brie-Comte-Robert (Paris). Honorary Member of the Academy of Sciences of the USSR (1923). Member of the Institute of France as a Foreign member of the Academy of Science, Foreign Member of the English Royal Society and a Honorary Member of the International Society of Soil Sciences (1924). Foreign Member of the Academy of Sciences of France (1925). Corresponding Member of the American Microbiological Society (1926). Honorary Member of the Biological Society of France (1929). Member of the Scientific Society of Italy (1931). Member of the Royal Academy of Science in Amsterdam (1932). Foreign Member of the Royal Academy of Science of Sweden (1934). Awarded the Leeuwenhoek Medal (1935). Honorary President of the IV International Botanical Congress in Amsterdam (1935). Honorary Member of the Society of Biological Chemists of India and the Society of American Bacteriologists (1936). Honorary Vice-President of the II International Microbiological Congress in London (1936). Awarded a Diploma of Ministry of New Jersey (1936). Honorary President of the International Botanical Congress in Stockholm (1940) and the IV International Microbiological Congress in Copenhagen (1947). Honorary Member of the Society of Soil Sciences of Florida (1950). Honorary Member of the Society of General Microbiologists of the UK (1952).
1953	Died in Brie-Comte-Robert.

This is not a comprehensive list of Winogradsky's research career or accomplishments.

Acknowledgments

We thank Dmitry Musolin, St. Petersburg State Forest Technical University and St. Petersburg State University, and Grozdilova Ludmila, Institute of Zoology of the Russian Academy of Sciences, St. Petersburg, for their assistance in our search of the literature. We also thank Dmitry Irodov and Oleksy Burlak for their assistance accessing some of the early publications and photography of the columns, and members of SIMBIOS who provided images and comments on the developing manuscript. Andrew Spiers is a member of SAGES (Scottish Alliance for Geoscience Environment and Society), and is supported by the SIMBIOS Centre and the University of Abertay Dundee. The University of Abertay Dundee is a charity registered in Scotland, No: SC016040.

References

[1] Ackert, L.T. (2006). The role of microbes in agriculture: Sergei Vinogradskii's discovery and investigation of chemosynthesis, 1880–1910. *J. History Biol.* 39: 373–406.

[2] Ackert, L.T. (2007). The ''Cycle of Life'' in ecology: Sergei Vinogradskii's soil microbiology, 1885–1940. *J. History Biol.* 40: 109–145.

[3] Ackert, L. (2013). Sergi Vinogradskii and the Cycle of Life. In: Archimedes 34: New Studies in the History and Philosophy of Science and Technology. Buchwald, J.Z. (Series Ed.), Springer, Dordrecht Heidelberg New York London.

[4] Acot, P. (1997). The Lamarckian cradle of scientific ecology. *Acta Biotheoretica* 45: 185–193.

[5] Amann, R. and Fuchs, B.M. (2008). Single-cell identification in microbial communities by improved fluorescence *in situ* hybridization techniques. *Nature Rev. Microbiol.* 6: 339 – 348.

[6] Anderson, D.C. and Hairston, R.V. (1999). The Winogradsky column and biofilms models for teaching nutrient cycling and succession in an ecosystem. *Amer. Biol. Teacher* 61: 453–459.

[7] Bantinaki, B., Kassen, R., Knight, C.G., Robinson, Z., Spiers, A.J. and Rainey, P.B. (2007). Adaptive divergence in experimental populations of *Pseudomonas fluorescens.* III. Mutational origins of Wrinkly Spreader diversity. *Genetics* 176: 441–453.

[8] Brockhurst, M.A., Habets, M.G.J.L., Libberton, B., Buckling, A. and Gardner, A. (2010). Ecological drivers of the evolution of public-goods cooperation in bacteria. *Ecology* 91: 334–340.

[9] Buckling, A., Kassen, R., Bell, G. and Rainey, P.B. (2000). Disturbance and diversity in experimental microcosms. *Nature* 408: 961 – 964.

[10] Buckling, A., Maclean, C.R., Brockhurst, M.A. and Colegrave, N. (2009). The Beagle in a bottle. *Nature* 457: 824–829.

[11] Canfield, D.E., Glazer, A.N. and Falkowski, P.G. (2010). The evolution and future of Earth's nitrogen cycle. *Science* 330: 192–196.

[12] Cholodny, N. (1930). Uber eine neue Methode zur Untersuchung der Bodenmikroflora. *Arch. Microbiol.* 1: 620–652.

[13] Cholodny, N.G. (1934). A soil chamber as a method for the microscopic study of the soil microflora. *Arch. Microbiol.* 5: 148–156.

[14] Crawford, J.W., Harris, J.A., Ritz, K. and Young, I.M. (2005). Towards an evolutionary ecology of life in soil. *TRENDS Ecol. Evol.* 20: 81–87.

[15] D'aes, J., De Meyer, K., Pauwelyn, E. and Höfte, M. (2010). Biosurfactants in plant-*Pseudomonas* interactions and their importance to biocontrol. *Environ. Microbiol. Reports* 2: 359-372.

[16] Downie, H., Holden, N., Otten, W., Spiers, A.J., Valentine, T.A. and Dupuy, L.X. (2012). Transparent soil for imaging the rhizosphere. *PLoS ONE* 7: e44276.

[17] Dworkin, M. (2012). Sergei Winogradsky: a founder of modern microbiology and the first microbial ecologist. *FEMS Microbiol. Rev.* 36: 364–379.

[18] Egunov, M.A. (1895). The sulfur bacteria of the Odessa Bay. *Arch. Sci. Biol. St. Petersburg* 3: 381–397.

[19] Eickhorst, T. and Tippkötter, R. (2008). Improved detection of soil microorganisms using fluorescence *in situ* hybridization (FISH) and catalyzed reporter deposition (CARD-FISH). *Soil Biol. Biochem.* 40: 1883–1891.

[20] Falkowski, P.G., Fenchel, T. and Delong, E.F. (2008). The microbial engines that drive Earth's biogeochemical cycles. *Science* 320: 1034–1039.

[21] Fechtner, J., Koza, A., Dello Sterpaio, P., Hapca, S.M. and Spiers, A.J. (2011). Surfactants expressed by soil pseudomonads alter local soil-water distribution suggesting a hydrological role for these compounds. *FEMS Microbiol. Ecol.* 78: 50–58.

[22] Francis, C.A., Beman, J.M. and Kuypers, M.M. (2007). New processes and players in the nitrogen cycle: the microbial ecology of anaerobic and archaeal ammonia oxidation. *ISME J.* 1: 19–27.

[23] Friedrich, C.G., Bardischewsky, F., Rother, D., Quentmeier, A. and Fischer, J. (2005). Prokaryotic sulfur oxidation. *Curr. Opin. Microbiol.* 8: 253–259.

[24] Frigaard, N.U. and Dahl, C. (2009). Sulfur metabolism in phototrophic sulfur bacteria. *Adv. Microb. Physiol.* 54: 103–200.

[25] Fuchsman, C.A., Kirkpatrick, J.B., Brazelton, W.J., Murray, J.W. and Staley, J.T. (2011). Metabolic strategies of free-living and aggregate-associated bacterial communities inferred from biologic and chemical profiles in the Black Sea suboxic zone. *FEMS Microbiol. Ecol.* 78: 586-603.

[26] Gal, M., Preston, G.M., Massey, R.C., Spiers, A.J. and Rainey, P.B. (2003). Genes encoding a cellulosic polymer contribute toward the ecological success of *Pseudomonas fluorescens* SBW25 on plant surfaces. *Mol. Ecol.* 12: 3109–3121.

[27] Giddens, S.R., Jackson, R.W., Moon, C.D., Jacobs, M.A., Zhang, X-X., Gehrig, S.M., Rainey, P.B. (2007). Mutational activation of niche-specific genes provides insight into regulatory networks and bacterial function in a complex environment. *Proc. Natl. Acad. Sci. (USA)* 104: 18247–18252.

[28] Ghosh, W. and Dam, D. (2009). Biochemistry and molecular biology of lithotrophic sulfuroxidation by taxonomically and ecologically diverse bacteria and archaea. *FEMS Microbiol. Rev.* 33: 999–1043.

[29] Green, J.H., Koza, A., Moshynets, O., Pajor, R., Ritchie, M.R. and Spiers, A.J. (2011). Evolution in a test-tube : rise of the Wrinkly Spreaders. *J. Biol. Educ.* 45: 54–59.

[30] Hapca, S.M., Wang, Z.X., Otten, W., Wilson, C. and Baveye, P.C. (2011). Automated statistical method to align 2D chemical maps with 3D X-ray computed micro-tomographic images of soils. *Geoderma* 164: 146–154.

[31] Huang, W.E., Bailey, M.J., Thompson, I.P., Whiteley, A.S. and Spiers, A.J. (2007). Single-cell Raman spectral profiles of *Pseudomonas fluorescens* SBW25 reflects *in vitro* and *in planta* metabolic history. *Microb. Ecol.* 53: 414–425.

[32] Huang, W.E., Ferguson, A., Singer, A.C., Lawson, K., Thompson, I.P., Kalin, R.M., Larkin, M.J., Bailey, M.J. and Whiteley, A.S. (2009). Resolving genetic functions within microbial populations: *in situ* analyses using rRNA and mRNA stable isotope probing coupled with single-cell Raman-fluorescence *in situ* hybridization. *Appl. Environ. Microbiol.* 75: 234–241.

[33] Huang, W.E. and Spiers, A.J. (2006). Consideration of future requirements for Raman microbiology as an examplar for the *ab initio* development of informatics frameworks for emergent OMIC technologies. *OMICS: J. Integ. Biol.* 10: 238–241.

[34] Huang, W.E., Ward, A.D. and Whiteley, A.S. (2009). Raman tweezers sorting of single microbial cells. *Environ. Microbiol. Reports* 1: 44–49.

[35] Huang, W.E., Stoecker, K., Griffiths, R., Newbold, L., Daims, H., Whiteley, A.S. and Wagner, M. (2007). Raman-FISH: combining stable-isotope Raman spectroscopy and fluorescence *in situ* hybridization for the single cell analysis of identity and function. *Environ. Microbiol.* 9: 1878–1889.

[36] Huang, W.E., Ude, S. and Spiers, A.J. (2007). *Pseudomonas fluorescens* SBW25 biofilm and planktonic cells have differentiable Raman spectral profiles. *Microb. Ecol.* 53: 471–474.

[37] Izquierdo, J.A. and Nüsslein, K. (2006). Distribution of extensive *nifH* gene diversity across physical soil microenvironments. *Microb. Ecol.* 51: 441–452.

[38] Jenal, U. and Malone, J. (2006). Mechanisms of cyclic-di-GMP signaling in bacteria. *Ann. Rev. Genet.* 40: 385–407.

[39] Johnson, D.B. and Hallberg, K.B. (2008). Carbon, iron and sulfur metabolism in acidophilic micro-organisms. *Adv. Microb. Physiol.* 54: 201–255.

[40] Kassen, R., Buckling, A., Bell, G. and Rainey, P.B. (2000). Diversity peaks at intermediate productivity in a laboratory microcosm. *Nature* 406: 508 – 512.

[41] Kassen, R. and Rainey, P.B. (2004). The ecology and genetics of microbial diversity. *Annu. Rev. Microbiol.* 58: 207–231.

[42] Kelly, D.P. (1971). Autotrophy: concepts of lithotrophic bacteria and their organic metabolism. *Annu. Rev. Microbiol.* 25: 177–210.

[43] King, E.O., Ward, M.K. and Raney, D.C. (1954). Two simple media for the demonstration of pyocyanin and fluorescin. *J. Lab. Clin. Med.* 44: 301–307.

[44] Koza, A. (2011). Adaptation and niche construction by *Pseudomonas fluorescens* SBW25. PhD thesis, University of Abertay Dundee, Dundee.

[45] Koza, A., Moshynets, O., Otten, W. and Spiers, A.J. (2011). Environmental modification and niche construction: developing O_2 gradients drive the evolution of the wrinkly spreader. *ISME J.* 5: 665–673.

[46] Lu, X., Al-Qadiri, H.M., Lin, M. and Rasco, B.A. (2011). Application of mid-infrared and raman spectroscopy to the study of bacteria. *Food Bioproc. Technol.* 4: 919–935.

[47] Madsen, E.L. (2011). Microorganisms and their roles in fundamental biogeochemical cycles. *Curr. Opin. Biotechnol.* 22: 456–464.

[48] Malone, J.G., Williams, R., Spiers, A.J. and Rainey, P.B. (2007). The structure-function relationship of WspR: a *Pseudomonas fluorescens* response-regulator with a GGDEF output domain. *Microbiology* 153: 980–994.

[49] Martínez-Espinosa, R.M., Cole, J.A., Richardson, D.J. and Watmough, N.J. (2011). Enzymology and ecology of the nitrogen cycle. *Biochem. Soc. Trans.* 39: 175–178.

[50] Moshynets, O.V., Koza, A., Dello Sterpaio, P., Kordium, V.A. and Spiers, A.J. (2011). Up-dating the Cholodny method using PET films to sample microbial communities in soil. *Biopol. Cell* 27: 199–205.

[51] Moshynets, O.V., Shpylova, S.P., Kordium, V.A. and Potters, G. (2012). Visualisation of endophytic bacteria in *Phyllostachys* sp. and *Fargesia* sp.. *Bamboo Sci. Culture* 25: 11–18.

[52] Moshynets, O.V., Shpylova, S.P., Spiers, A.J. and Kosakivska, I.V. (2010) The phytosphere of *Brassica napus* L.: a niche for *Pseudomonas fluorescens* SBW25. *Rep. Acad. Sci. Ukraine* 12: 150–153.

[53] Nielsen, L., Li, X. and Halverson, L.J. (2011). Cell-cell and cell-surface interactions mediated by cellulose and a novel exopolysaccharide contribute to *Pseudomonas putida* biofilm formation and fitness under water-limiting conditions. *Environ. Microbiol.* 13: 1342–1356.

[54] Nunan, N., Ritz, K., Rivers, M., Feeney, D.S. and Young, I.M. (2006). Investigating microbial micro-habitat structure using X-ray computed tomography. *Geoderma* 133: 398–407.

[55] Nunan, N., Wu, K., Young, I.M., Crawford, J.W. and Ritz, K. (2002). *In situ* patterns of soil bacterial populations, mapped at multiple scales, in an arable soil. *Microb. Ecol.* 44: 296–305.

[56] Nunan, N., Wu, K., Young, I.M., Crawford, J.W. and Ritz, K. (2006). Spatial distribution of bacterial communities and their relationships with the micro-architecture of soil. *FEMS Microbiol. Ecol.* 44: 203–215.

[57] Omelyansky, V.L. (1909). Osnovy mikrobiologii (The basics of microbiology). Publishing house of Imperial Academy of Sciences, St. Petersburg.

[58] Otten, W., Grinev, D., Baveye, P., Wang, Z., Hapca, S. and Wilson, C. (2010). Integrating physical and chemical techniques to characterise soil microsites. In: Gilkes, R.J. and Prakongkep, N. (Eds). Proceedings of the 19th World Congress of Soil Science; Soil Solutions for a Changing World. International Union of Soil Sciences, Brisbane.

[59] Otten, W., Pajor, R., Schmidt, S., Baveye, P.C., Hague, R. and Falconer, R.E. (2012). Combining X-ray CT and 3D printing technology to produce microcosms with replicable, complex pore geometries. *Soil Biol. Biochem.* 51: 53e55.

[60] Pajor, R., Falconer, R., Hapca, S. and Otten, W. (2010). Modelling and quantifying the effect of heterogeneity in soil physical conditions on fungal growth. *Biogeosci. Discuss.* 7: 3477–3501.

[61] Pasteur, L. (1868). Études sur le vinaigre, sa fabrication, ses maladies, moyens de les prévenir: nouvelles observations sur la conservation des vins par la chaleur. Gauthier-Villars, Paris.

[62] Paterson, S., Vogwill, T., Buckling, A., Benmayor, R., Spiers, A.J., Thomson, N.R., Quail, M., Smith, F., Walker, D., Libberton, B., Fenton, A., Hall, N. and Brockhurst,

M.A. (2010). Antagonistic coevolution accelerates molecular evolution. *Nature* 464: 275-278.

[63] Pearsall, J. (Ed.) (1998). The New Oxford Dictionary of English. Clarendon Press, Oxford.

[64] Peix, A., Ramírez-Bahena, M-H. and Velázquez, E. (2009). Historical evolution and current status of the taxonomy of genus *Pseudomonas*. *Infec. Genet. Evol.* 9: 1132–1147.

[65] Perfil'ev, B.V. and Gabe, D.R. (1969). Capillary methods of investigating micro-organisms. Oliver and Boyd, Edinburgh.

[66] Preston, G.M., Bertrand, N. and Rainey, P.B. (2001). Type III secretion in plant growth-promoting *Pseudomonas fluorescens* SBW25. *Mol. Microbiol.* 41: 999–1014.

[67] Quilès, F., Humbert, F. and Delille. A. (2010). Analysis of changes in attenuated total reflection FTIR fingerprints of *Pseudomonas fluorescens* from planktonic state to nascent biofilm state. *Spectrochimica Acta A* 75: 610–616.

[68] Raaijmakers, J.M., De Bruijn, I., Nybroe, O. and Ongena, M. (2010). Natural functions of lipopeptides from *Bacillus* and *Pseudomonas*: more than surfactants and antibiotics. *FEMS Microbiol. Rev.* 34: 1037–1062.

[69] Rainey, P.B. (1999). Adaptation of *Pseudomonas fluorescens* to the plant rhizosphere. *Environ. Microbiol.* 1: 243–257.

[70] Rainey, P.B., Buckling, A., Kassen, R. and Travisano, M. (2000). The emergence and maintenance of diversity: insights from experimental bacterial populations. *Trends Ecol. Evol.* 15: 243–247.

[71] Rainey, P.B. and Bailey, M.J. (1996). Physical and genetic map of the *Pseudomonas fluorescens* SBW25 chromosome. *Mol. Microbiol.* 19: 521–533.

[72] Rainey, P.B. and Travisano, M. (1998). Adaptive radiation in a heterogeneous environment. *Nature* 394: 69 –72.

[73] Robertson, M., Hapca, S.M., Moshynets, O. and Spiers, A.J. (2013). Air-liquid interface biofilm formation by psychrotrophic pseudomonads recovered from spoilt meat. *Ant. van Leeuwenhoek* 103: 251–259.

[74] Rogan, B., Lemke, M., Levandowsky, M. and Gorrell, T. (2005). Exploring the sulfur nutrient cycle using the Winogradsky column. *Amer. Biol. Teacher* 67: 348–356.

[75] Roslytsky, I. (2008). Sergiy Winogradsky – peredovyi mikrobioloh svity (Sergei Winogradskii is a world foremost microbiologist). Kyiv. Academperiodika, Kyiv.

[76] Schimel, J.P. and Schaeffer, S.M. (2012). Microbial control over carbon cycling in soil. *Front. Microbiol.* 3: 348.

[77] Schlegel, H.G. (1993). General microbiology. 7th Ed. Cambridge University Press, Cambridge.

[78] Schlegel, H.G. (1996). Winogradsky discovered a new *modus vivendi*. *Anaerobe* 2: 129-136.

[79] Schlegel, H.G. and Jannasch, H.W. (2006). Prokaryotes and their habitats. In: The Prokaryotes, Vol. 1: Symbiotic associations, Biotechnology, Applied Microbiology. 3rd Ed.. Dworkin, M., Falkow, S., Rosenberg, E., Schleifer, K.-H. and Stackebrandt, E. (Eds.). Springer, New York.

[80] Schmidt, S., Bengough, A.G., Gregory, P.J., Grinev, D.V. and Otten, W. (2012). Estimating root–soil contact from 3D X-ray microtomographs. *Eur. J. Soil Sci.* 63: 776–786.

[81] Schmidt, H., Eickhorst, T. and Tippkötter, R. (2012). Evaluation of tyramide solutions for an improved detection and enumeration of single microbial cells in soil by CARD-FISH. *J. Microbiol. Meth.* 91: 399–405.

[82] Seliber, G.L. (1960). Russkie mikrobiologi S.N. Vinogradskii i V.L. Omelianskii. Ministry of Agriculture of SSSR Press, Moscow.

[83] Silby, M.W., Cerdeño-Tárraga, A.M., Vernikos, G., Giddens, S.R., Jackson, R., Preston, G., Zhang, X-X., Godfrey, S., Spiers, A.J., Harris, S., Challis, G.L., Morningstar, A., Harris, D., Seeger, K., Murphy, L., Rutter, S., Squares, R., Quail, M.A., Saunders, E., Anderson, I., Mavromat, K., Brettin, T.S., Bentley, S., Thomas, C.M., Parkhill, J., Levy, S.B., Rainey, P.B. and Thomson, N.R. (2009). Genomic and functional analyses of diversity and plant interactions of *Pseudomonas fluorescens*. *Genome Biol.* 10: R51.

[84] Silby, M.W., Winstanley, C., Godfrey, S.A.C., Levy, S.B. and Jackson, R.W. (2011). *Pseudomonas* genomes: diverse and adaptable. *FEMS Microbiol. Rev.* 35: 652–680.

[85] Simon, J. and Klotz, M.G. (2013). Diversity and evolution of bioenergetic systems involved in microbial nitrogen compound transformations. *Biochim. Biophys. Acta.* 1827: 114–135.

[86] Skinner, C.E. and Fletcher, D.W. (1960). A review of the genus Candida. *Bacteriol. Rev.* 24: 397–416.

[87] Sorokin, Y.I. (1972). The bacterial population and the processes of hydrogen sulphide oxidation in the Black Sea. *J. Cons. Int. Explor. Mer.* 34: 423–454.

[88] Spiers, A.J. (2007). Wrinkly-spreader fitness in the two-dimensional agar plate microcosm: maladaptation, compensation and ecological success. *PLoS ONE* 2: e740.

[89] Spiers, A.J., Arnold, D.L., Moon, C.D. and Timms-Wilson, T.M. (2006). A survey of A-L biofilm formation and cellulose expression amongst soil and plant-associated *Pseudomonas* isolates. In: Microbial Ecology of Aerial Plant Surfaces. Bailey, M.J., Lilley, A.K. and Timms-Wilson, T.M. (Eds). CABI, Wallingford.

[90] Spiers, A.J., Buckling, A. and Rainey, P. (2000). The causes of *Pseudomonas* diversity. *Microbiology* 146: 2345–2350.

[91] Spiers, A.J., Deeni, Y.Y., Folorunso, A.O., Koza, A., Moshynets, O. and Zawadzki, K. (2013). Cellulose expression in *Pseudomonas fluorescens* SBW25 and other environmental pseudomonads. In: Cellulose. Van De Ven, T.G.M. and Trdine, J. (Eds). InTech, Rijeka.

[92] Spiers, A.J., Kahn, S.G., Bohannon, J., Travisano, M. and Rainey, P.B. (2002). Adaptive divergence in experimental populations of *Pseudomonas fluorescens*. I. Genetic and phenotypic bases of Wrinkly Spreader fitness. *Genetics* 161: 33–46.

[93] Spiers, A.J., Bohannon, J., Gehrig, S.M. and Rainey, P.B. (2003). Biofilm formation at the air–liquid interface by the *Pseudomonas fluorescens* SBW25 Wrinkly Spreader requires an acetylated form of cellulose. *Mol. Microbiol.* 50: 15–27.

[94] Spiers, A.J. and Rainey, P.B. (2005). The *Pseudomonas fluorescens* SBW25 Wrinkly Spreader biofilm requires attachment factor, cellulose fibre and LPS interactions to maintain strength and integrity. *Microbiology* 151: 2829–2839.

[95] Sternberg, C., Christensen, B.B., Johansen, T., Neilsen, A.T., Andersen, J.B., Givskov, M. and Molin, S. (1999). Distribution of bacterial growth activity in flow-chamber biofilms. *Appl. Environ. Microbiol.* 65: 4108-4117.

[96] Stewart, F.J. (2011). Dissimilatory sulfur cycling in oxygen minimum zones: an emerging metagenomics perspective. *Biochem. Soc. Trans.* 2011 39: 1859–1863.

[97] Taina, I.A., Heck, R.J. and Elliot, T.R. (2008). Application of X-ray computed tomography to soil science: A literature review. *Canadian J. Soil Sci.* 88: 1-19.

[98] Teske, A. and Nelson, D.C. (2006). The genera *Beggiatoa* and *Thioploca*. In: The Prokaryotes, Vol. 6: Proteobacteria: Gamma subclass. 3rd Ed.. Dworkin, M., Falkow, S., Rosenberg, E., Schleifer, K.-H. and Stackebrandt, E. (Eds.) Springer, New York.

[99] Tippkötter, R., Eickhorst, T., Taubner, H., Gredner, B. and Rademaker, G. (2009). Detection of soil water in macropores of undisturbed soil using microfocus X-ray tube computerized tomography (μCT). *Soil Tillage Res.* 105: 12–20.

[100] Tracy, S.R., Roberts, J.A., Black, C.R., McNeill, A., Davidson, R. and Mooney, S.J. (2010). The X-factor: visualizing undisturbed root architecture in soils using X-ray computed tomography. *J. Exp. Bot.* 61: 311–313.

[101] Ude, S., Arnold, D.L., Moon, C.D., Timms-Wilson, T. and Spiers, A.J. (2006). Biofilm formation and cellulose expression among diverse environmental *Pseudomonas* isolates. *Environ. Microbiol.* 8: 1997–2011.

[102] Ude, S., Bailey, M.J., Huang, W.E. and Spiers, A.J. (2007). The environmental plasmid pQBR103 alters the single-cell Raman spectral profile of *Pseudomonas fluorescens* SBW25. *Microb. Ecol.* 53: 494–497.

[103] Verschuren, P.G., Cardona, T.D., Nout, M.H.R., De Gooijer, K.D., Van den Heuvel, J.C. (2000). Location and limitation of cellulose production by *Acetobacter xylinum* established from oxygen profiles. *J. Biosci. Bioeng.* 89: 414–419.

[104] Waksman, S. (1953). Sergei N. Winogradsky: His Life and Work. Rutgers University Press, New Brunswick.

[105] Winogradsky, S. (1949). Microbiologie du sol, problèmes et méthodes; cinquante ans de recherches. (Soil Microbiology: Problems and Methods, Fifty Years of Investigations). Masson, Paris. (This book was also published elsewhere, for example, see the following reference for the Russian publication.)

[106] Winogradsky S.N (1952). Microbiologija pochvy. Problemy i metody. Pjatdesjat let issledovaniy (Soil Microbiology. Problems and Methods. Fifty Years of Investigations). Publishing house of the Academy of Science of USSR, Moscow.

[107] Xie, C., Mace, J., Dinno, M.A., Li, Y.Q., Tang, W., Newton, R.J. and Gemperline, P.J. (2005). Identification of single bacterial cells in aqueous solution using confocal laser tweezers Raman spectroscopy. *Anal. Chem.* 77: 4390–4397.

[108] Young, I.M., Crawford, J.W., Nunan, N., Otten, W. and Spiers, A. (2008) Microbial distribution in soils: physics and scaling. In: Advances in Agronomy. Vol. 100. Sparks, D.L. (Ed). Academic Press, Burlington.

[109] Zavarzin, G.A. (2006). Winogradsky and modern microbiology. *Microbiology* (ISSN 0026-2617) 75: 501–511.

[110] Zavarzin, G.A. (2006). Genius of natural science. On the 150th anniversary of the birth of honorary member of the USSR Academy of Sciences S.N. Vinogradskii. *Herald Russ. Acad. Sci.* (ISSN 1019-3316) 76: 361–369.

In: Microcosms
Editor: Christopher C. Harris

ISBN: 978-1-62618-661-3

Chapter 2

Soil Heterogeneity Modulates Responses to Multiple Global Environmental Changes in Model Grassland Communities

Fernando T. Maestre[1,2] and James F. Reynolds[2,3]

[1]Área de Biodiversidad y Conservación, Departamento de Biología y Geología, Escuela Superior de Ciencias Experimentales y Tecnología, Universidad Rey Juan Carlos, Móstoles, Spain
[2]Department of Biology, Duke University, Durham, NC, US
[3]Nicholas School of the Environment, Division of Environmental Science & Policy, Duke University, Durham, NC, US

Abstract

An important part of ecological research has been devoted to understand and predict how plant species and assemblages respond to changes in the availability and spatio-temporal heterogeneity of resources like water, light and nutrients. In the natural world, spatial heterogeneity in the availability of soil-based resources at various scales (soil heterogeneity) is the norm, rather than the exception, in most ecosystems. Albeit soil heterogeneity is a key feature of most terrestrial ecosystems, and plays a crucial role in shaping their structure, productivity, composition and diversity, it has often been ignored when evaluating the effects of global change drivers on plant performance and associated ecosystem processes. Thus, it is virtually unknown whether observed plant responses to global change drivers are modified by soil heterogeneity. This chapter synthesizes a series of multifactorial microcosm experiments aiming to evaluate the joint effects of soil heterogeneity and global change drivers (biotic diversity, nutrient availability, [CO_2] and changes in rainfall patterns) on the response of individuals and assemblages (in terms of biomass production and allocation and nutrient uptake patterns) formed by different combinations of *Lolium perenne* L., *Plantago lanceolata* L., *Trifolium repens* L., *Anthoxanthum odoratum* L. and *Holcus lanatus* L. We tested the hypothesis that soil heterogeneity modulates the magnitude of the response of both individuals and assemblages to multiple global change drivers (i.e., that individual and assemblage responses to soil heterogeneity and global change drivers will not be predictable from the observed responses to any one of the evaluated factors in isolation). In all the experiments conducted we did find a series of two- and three-term interactions determining key responses at both the individual and assemblage level. Thus, our working hypothesis was supported by our results. Our synthesis strongly suggest that single-factor studies should not be used to extrapolate how grassland ecosystems may respond to global change, which is composed of multiple—and potentially interacting—drivers.

INTRODUCTION

Changes in the atmospheric concentration of carbon dioxide ([CO_2]), nutrient availability and biotic diversity, as well as shifts in rainfall patterns towards larger, less frequent, rain events, are major drivers of the ongoing global environmental change being faced by terrestrial ecosystems worldwide (Vitousek et al. 1997; Sala et al. 2000; Easterling et al. 2000). Increasing concern with respect to the ecological consequences of such change has boosted research on the topic, and nowadays there is a plethora of studies evaluating the effects of its drivers on key ecosystem components, such as plants and soil organisms, and processes, such as net primary productivity and nutrient cycling (e.g., Bazzaz 1990; Körner & Bazzaz 1996; Poorter & Navas, 2003; Loreau et al. 2002; Fay et al. 2003; Niklaus & Körner 2004). Most of the research conducted to date on the topic has evaluated the ecological effects of the different global change drivers in isolation from each other (Strain & Cure 1994, Luo et al. 1999; Körner 2001; Loureau et al. 2002). During the last 15 years, however, global change research has developed towards more complex experiments that evaluate the joint effects of several drivers on organism performance and ecosystem processes, such as [CO_2], nutrient availability and biotic diversity (He et al. 2002; Reich et al. 2001, 2004), [CO_2], nutrient availability, temperature and rainfall (Shaw et al. 2002; Dukes et al. 2005), and [CO_2], temperature, rainfall and ozone (Hanson et al. 2005). These complex experiments have demonstrated that different drivers act simultaneously, and often interactively (but see Zavaleta et al. 2003), to determine the responses of ecosystem components and processes to global change, emphasizing the need of multifactorial approaches when evaluating them (Norby & Luo 2004, Kardol 2012).

Over the decades, an important part of ecological research has been devoted to understand and predict how plant species and assemblages respond to changes in the availability and spatio-temporal variability (hereafter named heterogeneity) of resources like water, light and nutrients (e.g., Grime 1979; Tilman 1988; Hutchings et al. 2000a). In the natural world, spatial heterogeneity in the availability of soil-based resources at various scales (hereafter termed soil heterogeneity) is the norm, rather than the exception, in most ecosystems (e.g., Jackson & Caldwell 1993; Ryel et al. 1996; Farley & Fitter 1999; Cain et al. 1999). Soil heterogeneity has profound consequences for plants, as it affects their establishment, neighborhood interactions, distribution, productivity and diversity (Stein et al. 1997; Pan et al. 1998; Robinson et al. 1999; Maestre et al. 2003; Désilets & Houle 2005). At spatial scales detectable by individual root systems, soil heterogeneity promotes a suite of plant physiological and morphological responses including changes in biomass allocation, root morphology, longevity and growth, and in nutrient uptake patterns (reviewed, among others, by Robinson 1994; Huber-Sanwald & Jackson 2001; Hutchings & John 2003; Hodge 2004; Hutchings et al. 2003). These responses determine the competitive ability and survival of individual plants within populations and assemblages (Day et al. 2003; Hodge 2004), as well as community composition and net primary productivity (Bliss et al. 2002; Wijeshinge et al. 2005; Maestre et al. 2006; but see Casper & Cahill 1996).

Albeit soil heterogeneity is a key feature of most terrestrial ecosystems, and plays a crucial role in shaping their structure, productivity, composition and diversity, it has often been ignored when evaluating the effects of global change drivers on plant performance and associated ecosystem processes (see García-Palacios et al. 2012 for a review). Thus, it is virtually unknown whether observed plant responses to global change drivers are modified by soil heterogeneity. The potential for them to be so is certainly great: soil heterogeneity is both ubiquitous and has been shown to have multiple effects on individual plant performance and plant-plant interactions (Hutchings et al. 2003; Hodge 2004). At the species level, interactions between global change drivers and soil heterogeneity may occur because the later promotes plant responses, such as changes in biomass allocation and nutrient uptake patterns, that are also a consequence of drivers such as elevated [CO_2] and increased nitrogen deposition (Bassirirad et al. 2001; Poorter & Navas 2003; Poorter & Nagel 2000; Hodge 2004). At the assemblage level, such interactions may occur because co-occurring plants often differ in their ability to profit from this heterogeneity (Bliss et al. 2002; Wijesinghe et al. 2005; García-Palacios et al. 2011) and in the direction and magnitude of their responses to global change drivers (Berntson et al. 1998; Grünzweig & Körner 2001).

Studies evaluating how plant species and assemblages respond to joint changes in nutrient heterogeneity and global change drivers are essential to advance in our understanding of the interactions between global change drivers and intrinsic ecosystem features, and to predict how such features may modify ecosystem responses to global change. To address the need of research in this area, we conducted a series of multifactorial microcosm experiments to evaluate the joint effects of soil heterogeneity and global change drivers (biotic diversity, nutrient availability, [CO_2] and changes in rainfall patterns) on the response of individuals and assemblages (in terms of biomass production and allocation and nutrient uptake patterns) formed by different combinations of *Lolium perenne* L., *Plantago lanceolata* L., *Trifolium repens* L., *Anthoxanthum odoratum* L. and *Holcus lanatus* L. These species commonly co-occur in semi-natural temperate grasslands of Europe (Joshi et al. 2000) and North America (Fowler & Antonovics 1981). In all the experiments, we used natural soil and organic material, rather than growing medium and inorganic fertilizer, to simulate the different soil heterogeneity and nutrient availability levels because they are relevant materials that permit interpretation of plant responses in a more realistic context (Hodge 2004). We tested the hypothesis that soil heterogeneity modulates the magnitude of the response of both individuals and assemblages to multiple global change drivers (i.e., that individual and assemblage responses to soil heterogeneity and global change drivers will not be predictable from the observed responses to any one of the evaluated factors in isolation). Interactions are predicted because both soil heterogeneity and the evaluated global change drivers modify productivity and biomass allocation patterns in grasslandcommunities (Wilsey & Potvin 2000; Wijesinghe et al. 2005; Niklaus & Körner 2004; García-Palacios et al. 2011), because non-additive responses may arise when two or more global change drivers act in combination (Reich et al. 2004), and because species-specific differences in traits related to their ability to respond to these environmental factors and soil heterogeneity (Grime & Mackey 2002; Poorter 1993; Kirkham et al. 1996; Clark et al. 1999; García-Palacios et al. 2011).

We used controlled environment facilites (CEFs) for this research because they allow to conduct complex multifactorial experiments at a reasonable cost, and to have adequate replication and environmental control, two key issues to adequately test null hypotheses, and to successfully tease out a signal (effect) against a background of noise (random variation). The use of experimental microcosms and CEFs in this study is not intended to mimic the full complexity of nature and to replace direct observation of ecosystem responses in manipulative field experiments, which are an important approach to the study of global change. Instead, microcosms permit us to elucidate interactions among the environmental factors evaluated that may occur in the field (Lawton 1995).

MATERIALS AND METHODS

Experimental Design

Four microcosm experiments were conducted at the Duke University Phytotron, located in Durham, North Carolina (USA). All of them shared the same microcosm design (Figure 1), but differed in the number of factors included in the design and in the number, density and identity of the species forming the assemblages. Microcosms consisted of PVC pipe (length 38 cm, internal diameter 10 cm) filled with, from the base, 5 cm of gravel (for drainage), and then 28 cm of a mixture of soil and sand (50:50, hereafter "background soil"). On top of the background soil we placed a 2 cm layer of a 50:50 mixture of organic Duke Forest soil: peat to avoid the formation of physical crusts in the surface of the microcosms, and to simulate the typical accumulation of organic matter in the topsoil of temperate grasslands (see Maestre et al. 2005 for details).

The first experiment (hereafter referred as Experiment 1) evaluated the effects of atmospheric [CO_2] (37.5 and 70 Pa), nutrient availability (40, 60 and 80 mg of N added as ^{15}N-labelled organic material), and soil heterogeneity (nutrients distributed homogeneously or heterogeneously, Figure 1) on assemblages formed by *Lolium*, *Plantago*, *Trifolium*, *Anthoxanthum* and *Holcus*. Two seedlings of each species, of similar size and developmental stage (two-leaf), were transplanted into each microcosm, resulting in a density of 1273 seedlings·m^{-2}. The planting pattern was allocated once at random and adhered to for all microcosms. We established 6 microcosms for each of the 12 treatment combinations, providing 72 assemblages total. These assemblages were placed in four walk-in growth chambers (two for each [CO_2]) within which climate was independently controlled. The [CO_2] control in Phytotron chambers is maintained using a centralized [CO_2] control system with specially designed hardware and software and LI-COR gas analyzers (LI-6262, LI-COR, Lincoln, Nebraska, USA). Chamber [CO_2] are sampled every 3-7 minutes and injection rates of pure CO_2 are controlled to maintain specific [CO_2] setpoints in the chambers. In general this system is able to maintain [CO_2] within 5% of setpoint. For each [CO_2] level, three replicates per treatment were randomly assigned to one of the chambers. The positions of the microcosms within each chamber were randomly interchanged every week. Temperatures in the growth chambers ranged from 12°C at night to 21°C during the day, and this regime included a simulated dawn and dusk period, each of two hours duration, where temperature was gradually ramped up or down. Relative humidity followed a similar pattern ranging from 85 to 70%, as did the lights. The maximum PAR in the 15 h photoperiod was 900 $\mu mol\ m^{-2}\ s^{-1}$. Microcosms were irrigated twice a day with 50 ml of distilled water during the first two weeks of the experiment and once a day with the same amount thereafter.

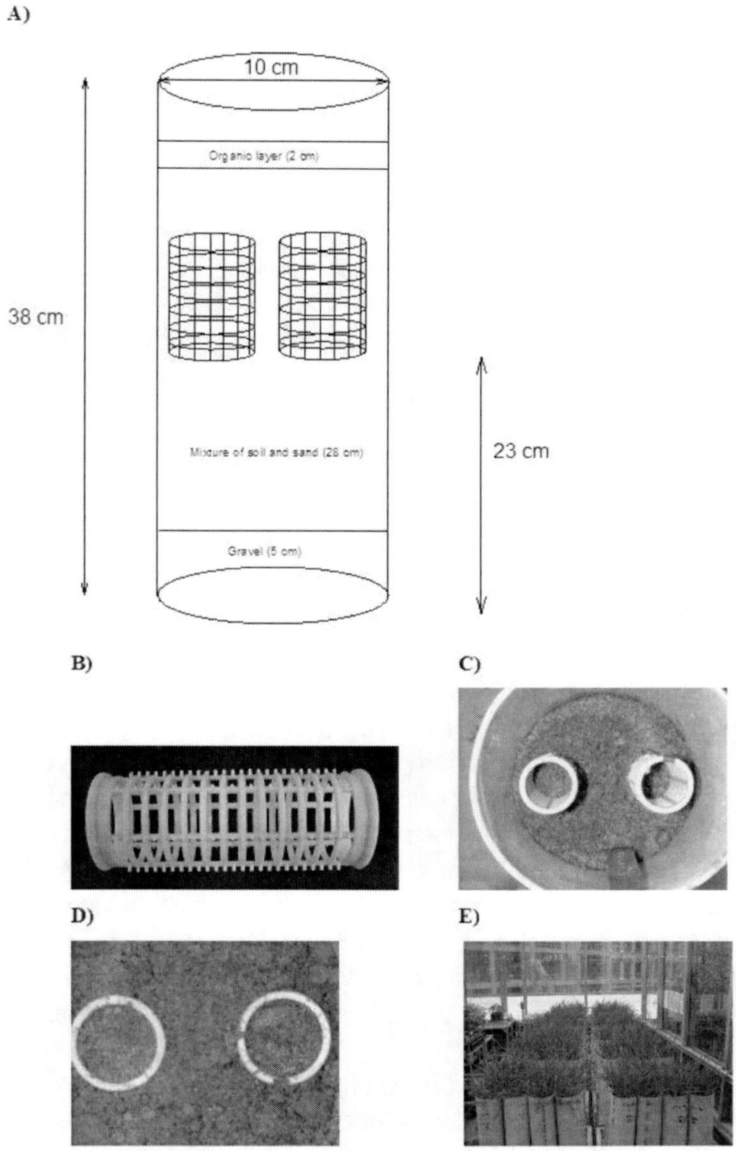

Reprinted with permission from *Ecology* (Maestre & Reynolds 2007b).

Figure 1. A) Schematic representation of the microcosms used (not drawn at scale). B) Detail of a plastic cylinder. C) View of the plastic cylinders placed in each microcosm. D) Detail of nutrient (right) and control (left) cylinders in a heterogeneous microcosm. E) View of the microcosms of Experiment 4 in the greenhouse after two months of growth.

The second experiment (hereafter referred as Experiment 2) evaluated the effects of atmospheric [CO_2] (37.5 and 70 Pa), soil heterogeneity (nutrients distributed homogeneously or heterogeneously) and plant species composition (*Lolium* monocultures, *Plantago* monocultures, *Holcus* monocultures, *Lolium* + *Plantago* mixtures, *Lolium* + *Holcus* mixtures, *Plantago* + *Holcus* mixtures and *Lolium* + *Plantago* + *Holcus* mixtures) on assemblage productivity and biomass allocation patterns. Such a design allowed us to evaluate the joint species of soil heterogeneity, [CO_2], species richness (1-3 species) and species composition. Seedlings of each species, of similar size and developmental stage (two-leaf), were transplanted into each microcosm as described above. Monocultures contained six seedlings of a single species, two-species mixtures three seedlings of each of two species, and three-species mixtures two seedlings of each species. In this way total plant density across microcosms was kept constant (764 seedlings·m^{-2}). We established 4 replicated microcosms for each of the 28 treatment combinations, resulting in 112 microcosms in total. The microcosms were placed in four walk-in growth chambers (two for each [CO_2]) as described above. For each [CO_2] level, half of the microcosms per treatment were randomly assigned to one of the chambers, and were then randomly grouped in two wheeled trolleys. The [CO_2] and trolleys were rotated between chambers every week, and the position of the microcosms within each chamber was randomized with every rotation. Temperatures in the growth chambers ranged from 12ºC at night to 21ºC during the day, and this regime included a simulated dawn and dusk period as described above. Relative humidity followed a similar pattern ranging from 85 to 70%, as did the lights. PAR was maintained at 500 µmol m^{-2} s^{-1} during the first week of the experiment, 750 µmol m^{-2} s^{-1} during the second week of the experiment, and at 1000 µmol m^{-2} s^{-1} thereafter. Each microcosm was irrigated daily with 30 mL of distilled water during the first two weeks of the experiment, and with 50 mL hereafter. To reduce limitations to plant growth due to low overall soil fertility, all the microcosms were watered with 50 mL of a nutrient solution containing 35 mg of Ca (added as $CaCl_2 \cdot 2H_20$) and 29 mg of Mg (added as $MgSO_4 \cdot 7H_2O$) twice during the course of the experiment.

The third experiment (hereafter referred as Experiment 3) evaluated the effects of atmospheric [CO_2] (37.5 and 70 Pa), soil heterogeneity (nutrients distributed homogeneously or heterogeneously), nutrient availability (40 and 120 mg of N added as organic material) and species evenness (1:1:1 or 4:1:1 ratio among species, varying the identity of the dominant species in the 4:1:1 mixtures) on the biomass and nutrient uptake patterns of assemblages formed by *Lolium*, *Plantago* and *Holcus*. Three seedlings of each species, of similar size and developmental stage (two-leaf), were transplanted into each microcosm as described above. We established 4 replicated microcosms for each of the treatment combinations, resulting in 128 microcosms in total.

The microcosms were introduced in four walk-in growth chambers (two for each [CO_2]) as described above. For each [CO_2], half of the microcosm units per treatment were randomly assigned to one of the chambers, and then were randomly grouped in five wheeled trolleys. The [CO_2] and trolleys were rotated between chambers every week, and the position of the microcosms within each chamber was randomized with every rotation. The environmental conditions within the chambers and the watering scheme using in this experiment were the same as described for Experiment 2.

The fourth experiment (hereafter referred as Experiment 4) evaluated the effects of nutrient availability (40, 80 and 120 mg of N added as organic material), soil heterogeneity (nutrients distributed homogeneously or heterogeneously), water availability (125, 250 and

375 mL of water added per week) and water heterogeneity (watering five times a week or watering once a week) on assemblages formed by *Lolium*, *Plantago*, *Trifolium*, *Anthoxanthum* and *Holcus*.

Seedlings (two of each species) were transplanted as in Experiment 1. We established 4 replicated microcosms for each of the 36 treatment combinations, resulting in 144 microcosms in total. The microcosms were introduced in a heated greenhouse exposed to natural sunlight (maximum photosynthetic active radiation = 1314.2 ± 389.6 μmol m^{-2} s^{-1} [mean ± SD, n = 100]) but within which temperature was controlled (23ºC during the day and 14ºC during the night). Average relative humidity ranged from 96% at night to 76% during the day.

The microcosms were arranged in four blocks. Each block contained a replicate per combination of treatments and was formed by 3 wheeled trolleys containing 12 microcosms each. The position of the blocks within the greenhouse was interchanged twice a week, and that of the microcosms within each block was randomized once a week to minimize position effects.

To establish soil heterogeneity and nutrient availability levels, different amounts of finely cut (2 mm) ^{15}N-labelled *Lolium* (Experiment 1; 2.91% N, 14.5 C: N) or unlabelled *Trifolium* (Experiments 2, 3 and 4; 3.9% N, 10.8 C:N) shoots were added to the microcosms. Within each nutrient availability level, the organic material was added homogeneously (homogeneous treatment) or as a patch (heterogeneous treatment). In the homogeneous treatment, the organic material was mixed with the background soil before introducing it into the PVC pipe.

In the heterogeneous treatment, the organic material was localized within discrete 31 cm^3 volumes of soil (Figure 1). To create one of these patches we mixed 25 cm^3 of background soil with the organic material and introduced the resulting mix into a 31 cm^3 plastic cylinder (length 75 mm; internal diameter 23 mm) consisting of a light mesh with square pores 5×10 mm in size. We refer to this as the patch cylinder. A second (control) cylinder, filled only with background soil, was placed 2 cm apart and parallel to the patch cylinder. Cylinders were located 12 cm below the surface of the organic soil. The homogeneous treatment received two plastic control cylinders only.

To generate the three levels of water availability in Experiment 4, we irrigated the microcosms with different amounts of distilled water: 125, 250 and 375 mL per week in the low, medium and high water availability levels, respectively (equivalent to 15.92, 31.84 and 47.76 $L{\cdot}m^{-2}$ of water per week). For each of these availability levels, the same amount of water was supplied as a single pulse applied once a week (pulse regime) or as smaller pulses applied five times a week (regular regime).

Thus, the amount of water added per event in the pulse and regular watering regimes were 125, 250 and 375 mL and 25, 50 and 75 mL, for the low, medium, and high water availability levels, respectively.

Measurements and Harvest

In all the experiments, plants were harvested after 90 – 100 days of growth, a period roughly equivalent to a full growing season. In all cases, the aboveground biomass of all the microcosms was cut at the soil surface and sorted by species. Leaves and stems were dried at

60ºC until constant weight. After these measurements, the soil was carefully removed from the microcosm unit and the roots were harvested without separating them by species. To measure root proliferation, we extracted the roots within each cylinder by cutting those outside it. The rest of the root system was also collected, and all the roots were dried as described above. Root proliferation was estimated in all cases with the index RII (Armas et al. 2004).

In the heterogeneous treatments, RII was calculated as (RB_p - RB_c)/(RB_p + RB_c), where RB_p and RB_c are the root biomass in the patch and control cylinders, respectively. RII ranges from –1 to +1: a value of zero indicates equal root growth in nutrient patches and background soil (no precision of foraging). Increasing positive values indicate increasing precision (i.e., root proliferation into the nutrient patch). In the homogeneous treatment, the control cylinder located in the same location as the patch cylinder in the heterogeneous treatment was treated as the patch cylinder for the purposes of calculating the RII index.

In the Experiment 1, we determined the uptake of nitrogen from the added organic material by each species. Shoots and leaves were ground to a fine powder and a sub-sample injected into an elemental analyser (Costech CHN Analyser, Milan, Italy) coupled to an isotope ratio mass spectrometer (Finnigan Delta Plus, Bremen, Germany) via a Finnigan Conflo Interface. The atom percent ^{15}N excess was calculated by subtracting 0.366 (atmospheric background). The amount of nitrogen captured was estimated as the percentage of the N added in the organic material that was captured by a species ([(mg ^{15}N in the foliar tissue) / (mg ^{15}N in the added organic material)]*100) or the assemblage ([(sum of the mg ^{15}N obtained in roots and in the leaves of each species) / (mg ^{15}N in the added organic material)]*100) (Hodge et al. 2000). In the Experiment 3, aboveground tissue samples from all the species were ground to a fine powder in a ball mill and injected into an elemental analyzer as described above to obtain their nitrogen concentration.

RESULTS

Responses at the Assemblage Level

In all the experiments, assemblages had higher total biomass when the nutrients were heterogeneously supplied (Figure 2). Soil heterogeneity also modulated the effects of species composition, nutrient availability, water heterogeneity and water availability on this variable, as indicated by a series of two- and three-way interactions (Tables 1 and 2). Detailed analyses of these interactions revealed that in the Experiment 2, total biomass increased under heterogeneous conditions in all cases, but in the two-species mixtures involving *Holcus* such an increase was only marginally significant (*Plantago-Holcus* and *Lolium-Holcus* mixtures, statistical analyses not shown). In Experiments 3 and 4, positive responses to soil heterogeneity were especially evident under high nutrient supply (Figures 2C and 2D). In the later experiment, the magnitude of the increase in biomass in response to water availability was higher when the nutrients where heterogeneously supplied.

Table 1. Overview of the effects of the factors evaluated on assemblage responses in Experiments 1, 2 and 4. Results were analyzed in all cases using appropriate analysis of variance (ANOVA) models. Ne = non-evaluated factor; • = no significant effect ($P > 0.05$); + = increased values of this factor led to increased response ($P < 0.05$); - = increased values of this factor led to reduced response ($P < 0.05$); X = significant main effect ($P < 0.05$) and interaction ($P < 0.05$) with other factor; x = no significant effect ($P > 0.05$) but interaction ($P < 0.05$) with other factor. In the later two cases, the factor/s involved in the interaction are denoted as subscripts. Two "X" in the same cell indicate that the factor is involved in two independent interactions. See the main text for details on the number of levels within each factor. TB = total biomass, BAR = below: aboveground biomass, [N] = aboveground nitrogen concentration, NC = proportion of the nitrogen added with the organic material captured by the assemblages, RII = root foraging precision (responses to factors other than soil heterogeneity are shown for the heterogeneous treatment only)

	Experiment 1					Experiment 2			Experiment 4		
	TB	BAR	RII	[N]	NC	TB	BAR	RII	TB	BAR	RII
CO_2 concentration (CO_2)	•	X_{na}	•	x_{sh}	•	•	•	•	Ne	Ne	Ne
Soil heterogeneity (sh)	+	•	+	x_{co2}	+	X_{sc}	•	+	$X_{wh,wa}$, X_{wa}	$X_{na,wa}$	+
Nutrient availability (na)	+	x_{co2}	+	•	•	Ne	Ne	Ne	X_{wa}	$X_{sh,wa}$, X_{wh}	+
Species richness (sr)	Ne	Ne	Ne	Ne	Ne	•	•	•	Ne	Ne	Ne
Species composition (sc)	Ne	Ne	Ne	Ne	Ne	X_{sh}	*	•	Ne	Ne	Ne
Water heterogeneity (wh)	Ne	Ne	Ne	Ne	Ne	Ne	Ne	Ne	x_{wa}, $x_{sh, wa}$	$X_{na,}$ X_{wa}	•
Water availability (wa)	Ne	Ne	Ne	Ne	Ne	Ne	Ne	Ne	$X_{wh, sh}$, $X_{wh,}$ X_{sh}	$X_{na, sh}$, X_{wh}	•

* This factor significantly affected total biomass, but its direction varied with the identity of the species forming the assemblages.

Table 2. Overview of the effects of the factors evaluated on the biomass and nutrient responses at the assemblage level in the Experiment 3. Results were analyzed in all cases using appropriate analysis of variance (ANOVA) models. Separate analyses were conducted for each of the species dominating the 4:1:1 mixtures. BAR = below: aboveground biomass ratio. Rest of legend as in Table 1

Species dominating the 4:1:1 mixture	Response variable	Total biomass	BAR	Root foraging precision	Aboveground N concentration
Plantago lanceolata	Species evenness (se)	•	•	•	•
	CO_2 concentration (CO_2)	X_{na}	X_{na}	•	-
	Soil heterogeneity (sh)	X_{na}	-	+	•
	Nutrient availability (na)	$X_{co2,}$ X_{sh}	X_{co2}	+	•
Lolium perenne	Species evenness (se)	+	-	•	•
	CO_2 concentration (CO_2)	X_{na}	-	•	-
	Soil heterogeneity (sh)	X_{na}	-	+	+
	Nutrient availability (na)	$X_{co2,}$ X_{sh}	-	+	+
Holcus lanatus	Species evenness (se)	-	•	•	+
	CO_2 concentration (CO_2)	X_{na}	-	•	-
	Soil heterogeneity (sh)	+	X_{na}	+	•
	Nutrient availability (na)	X_{co2}	X_{nh}	+	•

Elevated [CO_2] promoted an increase in total assemblage biomass in the Experiment 3, albeit this effect was only observed under high nutrient supply (Figure 2C). Total biomass also increased with the number of species included in the assemblages (Experiment 2, Figure 2B), albeit this increase was not significant in any case (Table 1). The effects of species evenness on total biomass varied from negative to positive, depending on the identity of the species dominating the mixtures (Experiment 3, Figure 2C, Table 2).

Most of the factors evaluated affected biomass allocation patterns (Tables 1 and 2). A significant [CO_2] × nutrient availability interaction was found when analyzing the below: aboveground biomass ratio (BAR) in the Experiment 1 (Table 1). Analyses conducted at each nutrient availability level revealed that elevated [CO_2] reduced the BAR of assemblages, but only under medium and high nutrient supply (Figure 3A).

In the Experiment 2, the BAR of assemblages was only affected by species composition (Figure 3B).

In the Experiment 3, the effects of the factors on this variable varied depending on the species dominating the 4:1:1 mixtures (Table 2). When *Plantago* was the dominant species, the assemblages had lower BAR under elevated [CO_2], albeit the magnitude of such response was higher under high nutrient supply (Figure 3C). When *Lolium* was the dominant species in the 4:1:1 mixtures, assemblages had lower BAR values under elevated [CO_2], maximum species evenness, homogeneous nutrient supply and high nutrient availability (Figure 3C).

When *Holcus* was the dominant species in the 4:1:1 mixtures, soil heterogeneity modulated the effects of [CO_2], as assemblages had lower BAR in response to elevated [CO_2] only when the nutrients were heterogeneously supplied. In the Experiment 4, significant NH × NA × WA, WH × WA and NA × WH interactions determined BAR responses (Table 1). Analyses conducted at each water availability level showed that, at the low and medium water availability levels, assemblages had lower BAR when the nutrients were heterogeneously supplied and when nutrient availability increased. Analyses conducted at each nutrient availability level showed that, at the medium and high nutrient availability levels, the BAR of assemblages decreased under heterogeneous nutrient supply. At the high nutrient availability level, the BAR of assemblages decreased under pulse watering, being this effect more pronounced at the low water availability level (Figure 3D).

When the organic material was supplied heterogeneously, the assemblages demonstrated highly precise root foraging patterns in all cases (Figure 4). Root foraging precision increased with increases in nutrient availability (Experiments 1, 3 and 4; Tables 1 and 2). In Experiment 4, higher root proliferation was found in the lower as compared to the medium and high water availability levels (Figure 4D).

In the Experiment 1, a significant [CO_2] × soil heterogeneity interaction was found for the aboveground nitrogen concentration (Table 1, Figure 5A). Under homogeneous nutrient supply, assemblages had lower nitrogen concentrations under elevated [CO_2]. Such an effect was not found under heterogeneous conditions (statistical analyses not shown). Assemblages captured more nitrogen from the organic material when it was added as a patch (Table 1, Figure 5B). In the Experiment 3, and regardless the species dominating the 4:1:1 mixtures, the nitrogen concentration in aboveground tissues decreased under elevated [CO_2] (Figure 5C, Table 2). When *Lolium* and *Holcus* were dominant, it increased under high nutrient supply and maximum species evenness, respectively.

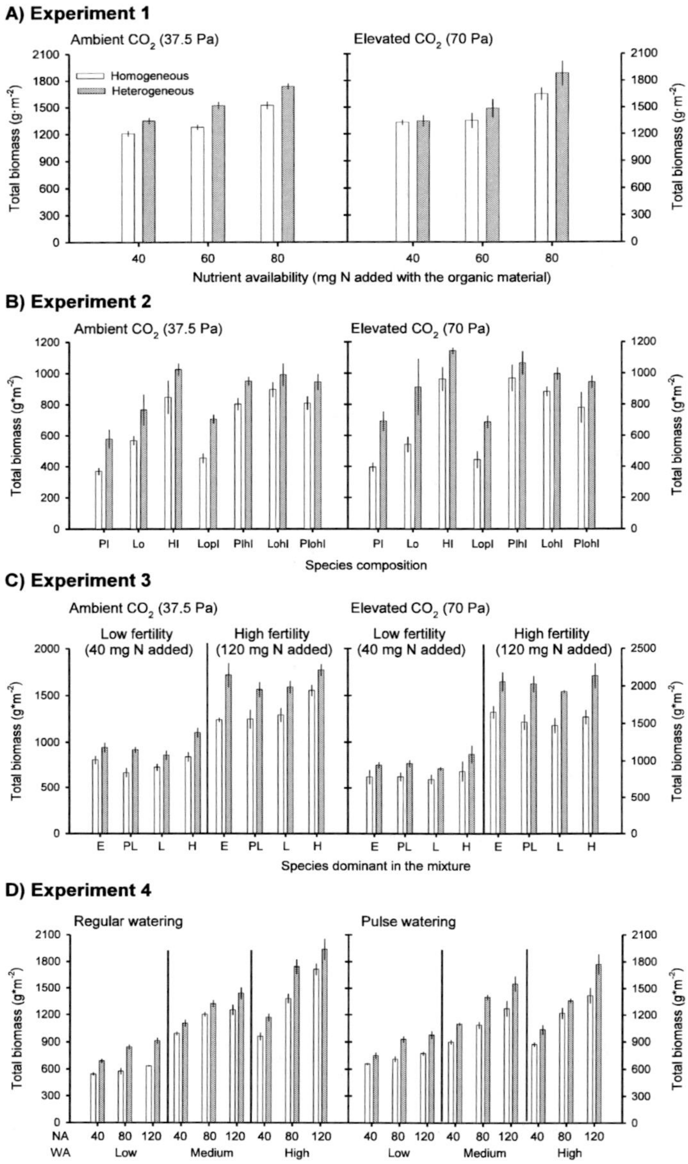

Redrawn with permission from the New Phytologistogist Trust (A: Maestre et al. 2005), *Oecologia* (B: Maestre & Reynolds 2007a), *Global Change Biology* (C: Maestre & Reynolds 2006b) and *Ecology* (D: Maestre & Reynolds 2007b).

Figure 2. Total assemblage biomass in the Experiments 1 (A), 2 (B), 3 (C) and 4 (D), compared across soil heterogeneity and global change drivers levels. Data are means ± SE ($n = 6$ for data in A, $n = 4$ for data in B, C, and D). Lo = *Lolium perenne* monoculture, Pl = *Plantago lanceolata* monoculture, Hl = *Holcus lanatus* monoculture, Lopl = *Lolium* + *Plantago* mixtures, Plhl = *Plantago* + *Holcus* mixtures, Lohl = *Lolium* + *Holcus* mixtures, Plohl = *Plantago* + *Lolium* + *Holcus* mixtures, E = equal evenness mixture (1 *Plantago*: 1 *Lolium* : 1 *Holcus*), P = mixture dominated by *Plantago lanceolata* (4 *Plantago*: 1 *Lolium* : 1 *Holcus*), L = mixture dominated by *Lolium perenne* (1 *Plantago*: 4 *Lolium* : 1 *Holcus*), H = mixture dominated by *Holcus lanatus* (1 *Plantago*: 4 *Lolium* : 1 *Holcus*), NA = nutrient availability and WA = water availability.

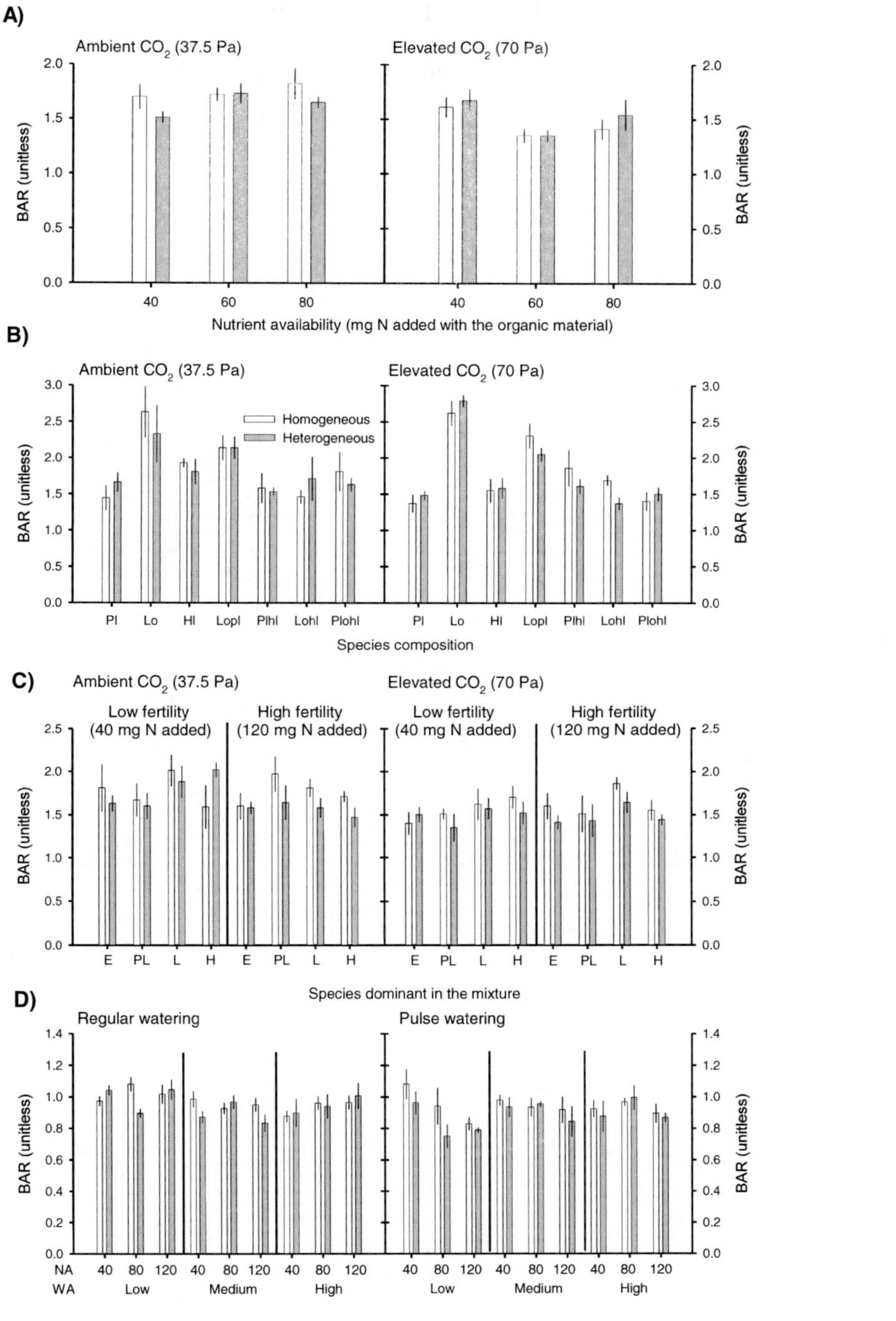

Redrawn with permission from the New Phytologistogist Trust (A: Maestre et al. 2005), *Oecologia* (B: Maestre & Reynolds 2007a), *Global Change Biology* (C: Maestre & Reynolds 2006b) and *Ecology* (D: Maestre & Reynolds 2007b).

Figure 3. Below: aboveground biomass ratio (BAR) of the assemblages in the Experiments 1 (A), 2 (B), 3 (C) and 4 (D), compared across soil heterogeneity and global change drivers levels. Data are means ± SE (n = 6 for data in A, n = 4 for data in B, C, and D). For explanation of labels see Figure 1 caption.

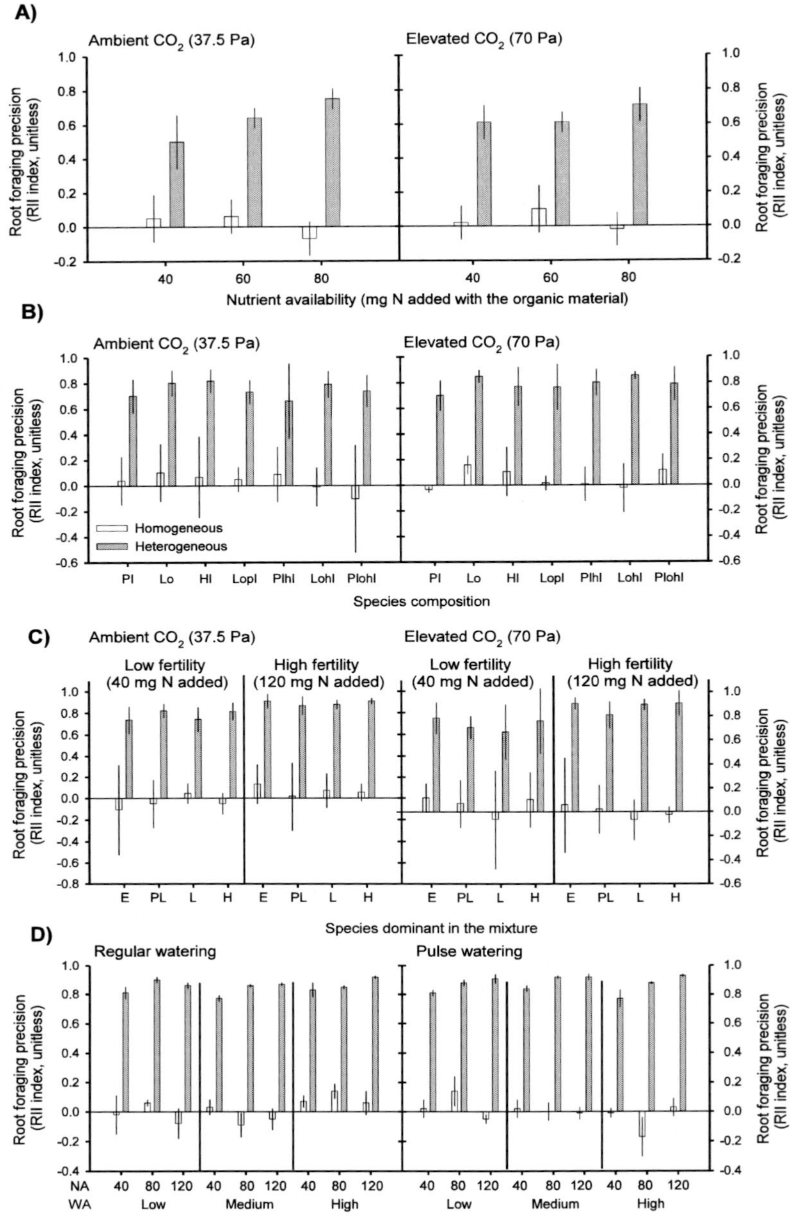

Redrawn with permission from the New Phytologistogist Trust (A: Maestre et al. 2005), Oecologia (B: Maestre & Reynolds 2007a), Global Change Biology (C: Maestre & Reynolds 2006b) and Ecology (D: Maestre & Reynolds 2007b).

Figure 4. Root foraging precision into nutrient patches (RII index) of the assemblages in the Experiments 1 (A), 2 (B), 3 (C) and 4 (D), compared across soil heterogeneity and global change drivers levels. Data are means ± 95% confidence intervals (n = 6 for data in A, n = 4 for data in B, C, and D). Positive values of the RII index indicate precise root foraging (proliferation) responses. Significant root proliferation is shown by confidence intervals that do not overlap zero. For explanation of labels see Figure 1 caption.

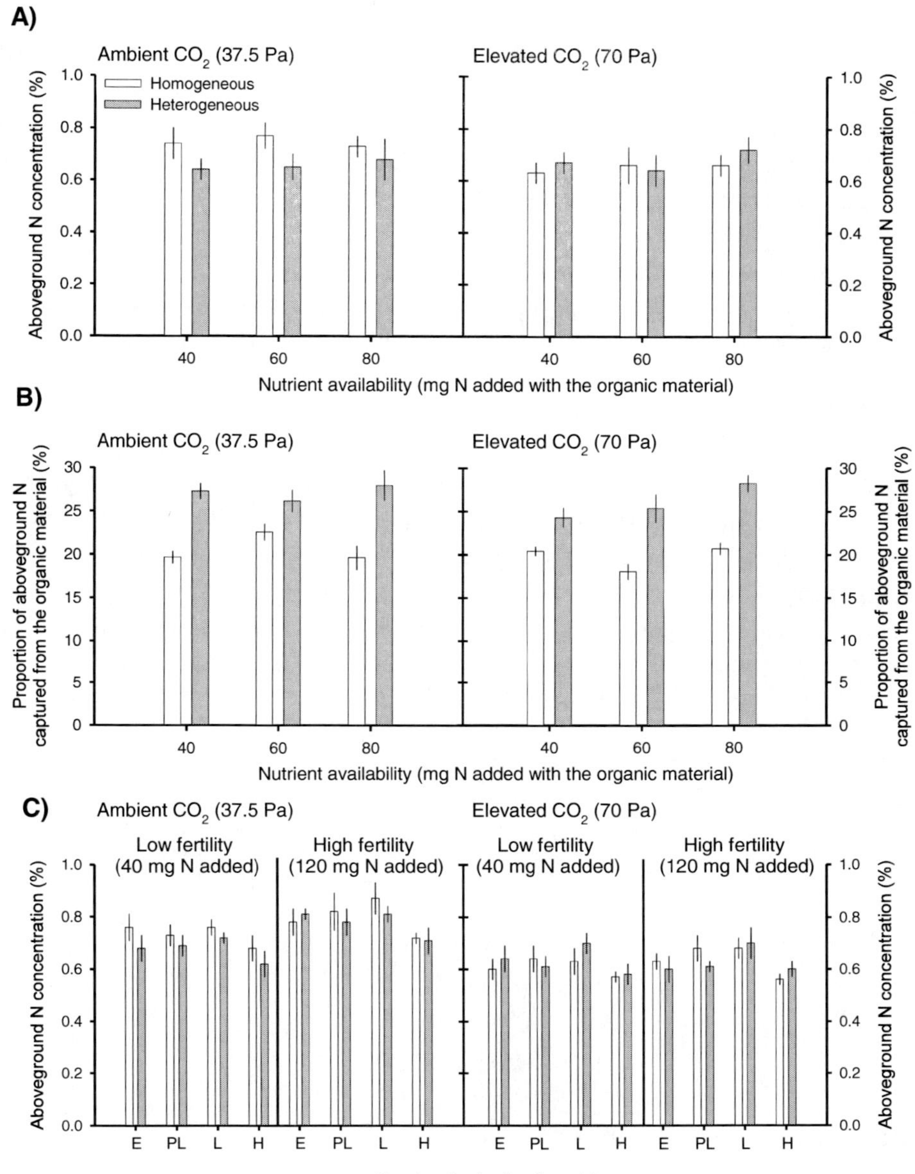

Redrawn with permission from the New Phytologistogist Trust (A and B: Maestre et al. 2005) and Global Change Biology (C: Maestre & Reynolds 2006b).

Figure 5. Aboveground nitrogen concentration (A) and proportion of nitrogen captured from the organic material added (B) in the assemblages of Experiment 1, and aboveground nitrogen concentration (C) in the assemblages of Experiment 3. Data are means ± SE (n = 6 for data in A and B, n = 4 for data in C). For explanation of x-axis labels see Figure 1 caption.

Responses at the Individual Level

Individual responses were evaluated in Experiments 1 and 4. In both experiments, *Holcus lanatus* was the dominant species in the assemblages: it accounted for >50% of the mean total aboveground biomass in all treatment combinations (Figures 6 and 7). *Lolium perenne* and *A. odoratum* were the next most dominant; they had proportions ranging from 7% to 31% and from 7% to 44% of the total aboveground biomass, respectively.

Biomass of *T. repens* and *P. lanceolata* were always below 18% and 8% of the aboveground total, respectively. In the Experiment 1, the proportions of the total aboveground biomass accounted for by *P. lanceolata* and *H. lanatus* were affected by a significant [CO_2] × soil heterogeneity interaction (Figure 6).

Investigation of these interactions revealed that, at ambient [CO_2], the proportion of aboveground biomass accounted for by *P. lanceolata* was lower under heterogeneous conditions of nutrient supply, but that accounted for by *H. lanatus* was greater (statistical analyses not shown). *Plantago lanceolata* had also lower proportions under the highest as compared to the lowest nutrient availability (Figure 6). In the Experiment 4, soil heterogeneity significantly decreased the proportion of the total aboveground biomass accounted for by *Plantago*, *Trifolium* and *Anthoxanthum*, but increased that of *Lolium* (Figure 7). The decrease in the proportion accounted for by *Anthoxanthum* under heterogeneous nutrient supply was only observed at the regular watering regime, as indicated by a significant soil heterogeneity × water heterogeneity interaction (Figure 7). The proportion of total aboveground biomass accounted for by *Trifolium* and *Plantago* was higher at the low compared to the medium and high nutrient availability levels, and at the pulse compared to the regular watering regime (statistical analyses not shown).

The proportion of total aboveground biomass accounted for by *Lolium* and *Plantago* was higher at the low compared to the medium and high nutrient availability levels, but that of *Holcus* was higher at the medium and high nutrient availability levels (compared to the low nutrient availability level, post-hoc results not shown).

In the Experiment 1, and when data from across all species were combined, the enhancement in nitrogen uptake at elevated relative to ambient [CO_2] was positively related to that in aboveground biomass production (Figure 8).

DISCUSSION

Responses at the Assemblage Level

In all the experiments, assemblage biomass increased when the organic material was heterogeneously supplied, a response observed in previous studies conducted with grassland populations and assemblages (Day et al. 2003; Wijesinghe et al. 2005; Maestre et al. 2006; Maestre & Reynolds 2006, García-Palacios et al. 2011; but see Casper & Cahill 1996, 1998). Soil heterogeneity also modulated the effects of (i) nutrient availability, water heterogeneity, water availability and species composition on biomass production, (ii) nutrient availability, water availability and [CO_2] on the below: aboveground biomass ratio, and (iii) [CO_2] on the aboveground nitrogen concentration (Tables 1 and 2). These results suggest that soil heterogeneity may be an extremely important environmental factor when assessing and interpreting the effects of global change drivers on plant assemblages (García-Palacios et al. 2012).

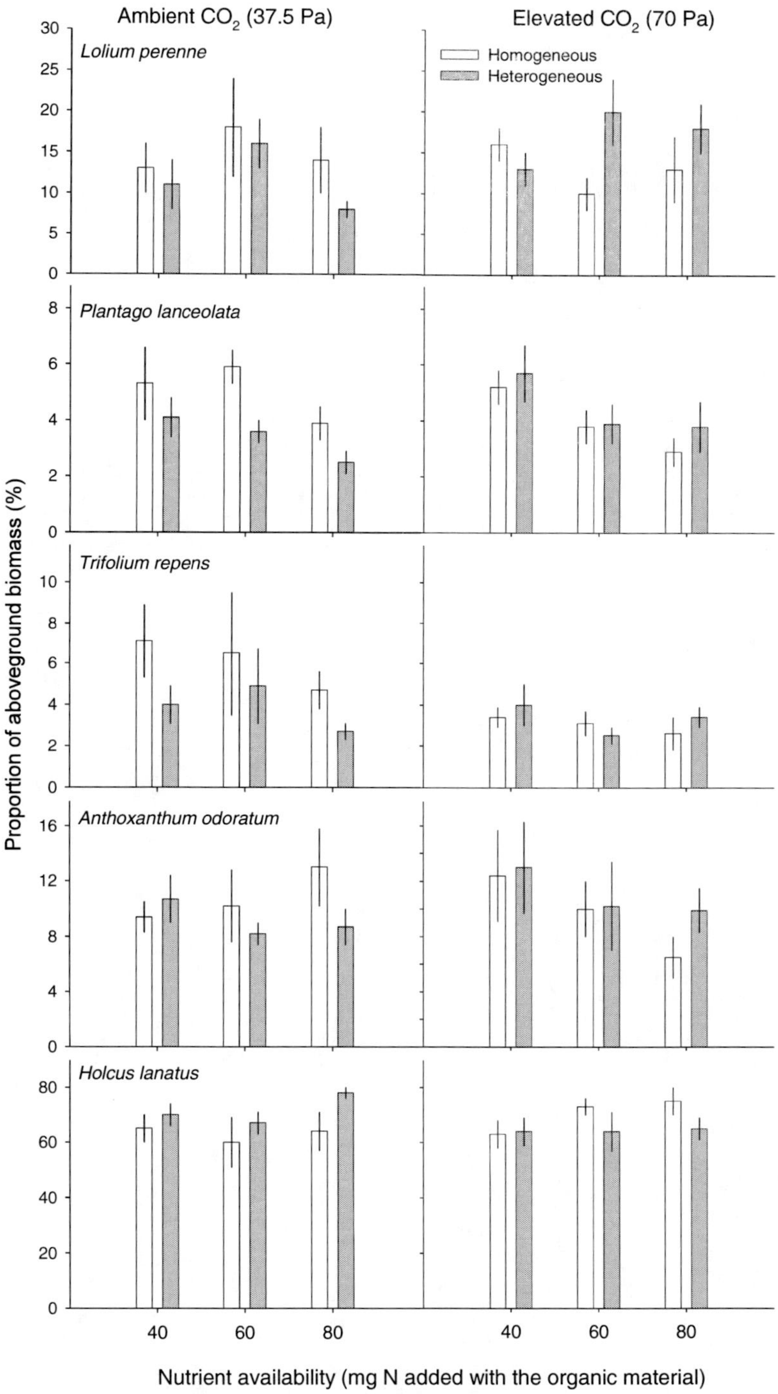

Redrawn with permission from the New Phytologistogist Trust (Maestre et al. 2005).

Figure 6. Proportion of the total aboveground biomass accounted for by each species in the assemblages (Experiment 1). Note the difference in scale of the y-axes. Data are means ± 1 SE (n = 6).

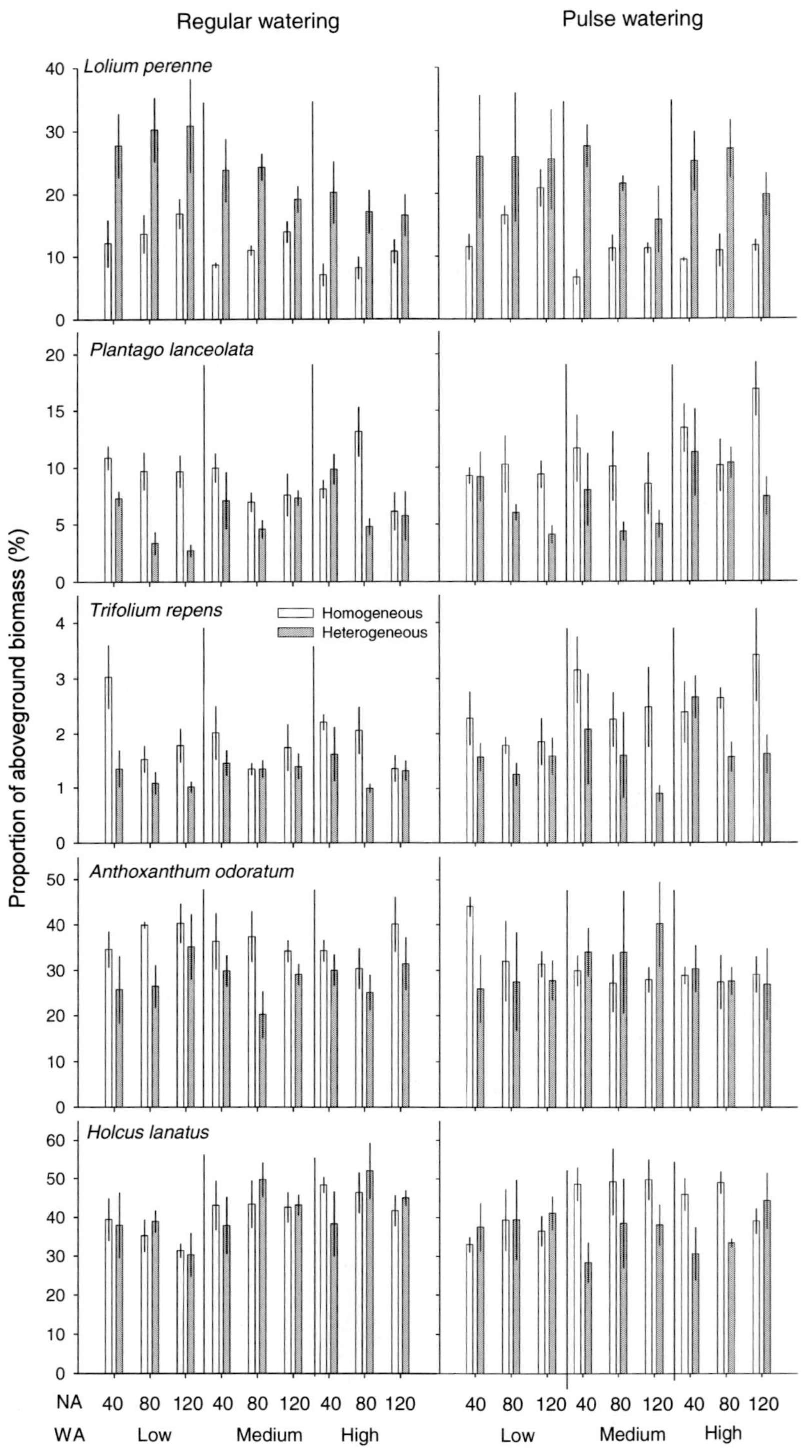

Redrawn with permission from Ecology (Maestre & Reynolds 2007b).

Figure 7. Proportion of the total aboveground biomass accounted for by each species in the assemblages (Experiment 4). Note the difference in scale of the y-axes. Data are means ± 1 SE (n = 4). NA = nutrient availability and WA = water availability.

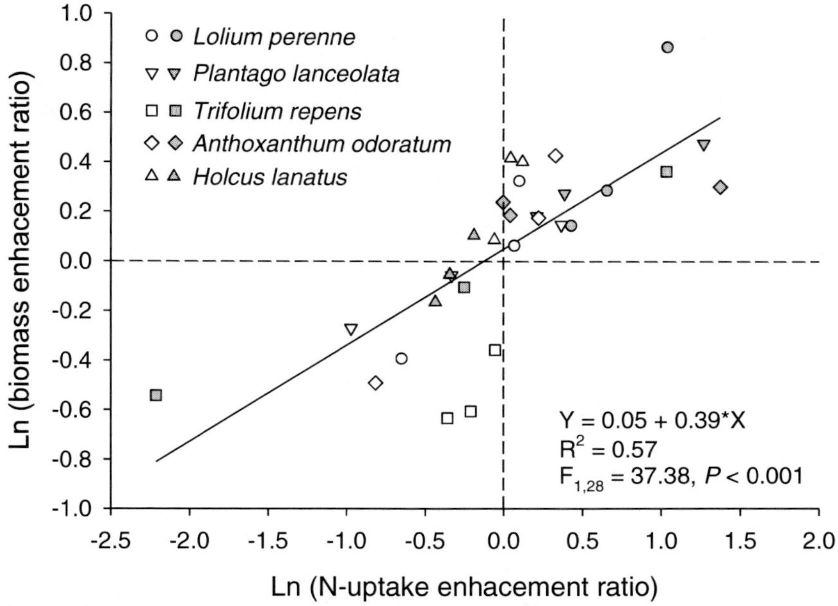

Redrawn with permission from the New Phytologistogist Trust (Maestre et al. 2005).

Figure 8. Relationship between the nitrogen uptake enhancement ratio (NER) and the biomass enhancement ratio (BER) for the different species in the Experiment 1. For a given species and combination of soil heterogeneity and nutrient availability treatments, BER and NER values were obtained by dividing the average aboveground biomass and proportion of nitrogen captured from the organic material added obtained at 70 Pa [CO_2] by those obtained at 37.5 Pa [CO_2], respectively. For clarity of presentation, the different nutrient availability levels are not shown. Grey symbols: heterogeneous treatments; White symbols: homogeneous treatments.

Our results suggest that the proliferation of roots into nutrient patches is an important driver of observed responses to soil heterogeneity. It has been shown that when a fixed amount of nutrients are made available to a plant, their acquisition will be more efficient if these are spatially concentrated because of preferential root allocation in these areas (Fransen et al. 1999). Individual plants begin to proliferate roots quickly after they encounter a nutrient patch, and this response increases nutrient uptake and overall plant growth (Hodge 2004). This assertion seems to be supported by the fact the assemblages captured more nitrogen from the added organic material added when it was supplied heterogeneously (Figure 3B), and their biomass increased linearly with increased foraging precision in Experiments 1, 2 and 3 (data not shown). Increases in plant growth, in turn, further stimulates root proliferation because of the positive relationship between plant size and/or relative growth rate and root foraging precision (Einsmann et al. 1999; Rajaniemi & Reynolds 2004; Kembel & Cahill 2005; Fransen et al. 1999), in a positive feedback that increases growth until the supply of nutrients is exhausted (Fransen et al. 2001). This positive feedback should be amplified under high nutrient availability conditions because of increases in plant size in response to the addition of nutrients. The increase in root foraging precision observed in response to this addition, and the positive relationships found between root proliferation into nutrient patches and total biomass support this view. It must be noted that root foraging responses to water availability did not fully match biomass responses, as biomass increased with increases in water availability but root foraging precision did not. These results suggest that the relative importance of root foraging versus other root responses elicited by nutrient patchiness, such as changes in nutrient uptake rates and root longevity (Hodge 2004), may depend on water availability.

As in previous studies conducted with grassland species (e.g. Stöcklin et al. 1998; He et al. 2002; Grünzweig & Körner 2003), we found significant interactions between [CO_2] and nutrient availability for assemblage attributes such as aboveground biomass, the below: aboveground ratio and the amount of nitrogen captured from the added organic material (Experiments 1 and 3). The increase in nutrient availability increased growth and nitrogen capture responses to elevated [CO_2], suggesting that [CO_2] effects on these assemblage attributes were constrained by low nutrient availability. As commonly observed with grassland species (Cotrufo et al. 1998; Bradford et al. 2012), elevated [CO_2] promoted a decrease in the aboveground nitrogen concentration. Similar nitrogen concentration values to those found here have been reported by previous experiments that have grown the same species using field soil as growing medium and plant material as fertilizer (Hodge et al. 2000; Wurst et al. 2003).

Among the biodiversity attributes evaluated, species richness was the only factor that did not influence productivity patterns. The lack of a significant [CO_2] × richness interaction on the productivity of the assemblages also contrasts with the results of previous studies. Using a wider range of richness levels than used here (between 1 and 12 species), Reich et al. (2001) and He et al. (2002) found that the enhancement of biomass accumulation in response to elevated levels of [CO_2] was mostly achieved in species-rich assemblages. However, Reich et al. (2004) did not find strong evidence for [CO_2] × species richness interactions when evaluating the productivity of grassland assemblages of varying diversity (from 1 to 16 species). These contrasting results are not fully unexpected, as responses to species richness are often overwhelmed by the effects of species composition or functional diversity (García-Palacios et al. 2011), and vary depending of abiotic factors such as nutrient availability (Hooper et al. 2005). The identity of the species forming the assemblages (Experiment 2), and that of the dominant species in the 4:1:1 mixtures (Experiment 3), strongly influenced biomass production and allocation patterns. In terms of biomass production, the species evaluated here have been shown to respond positively to elevated [CO_2] (Poorter, 1993), nutrient heterogeneity (Grime & Mackey, 2002) and increases in nutrient availability (Kirkham et al. 1996). However, they differ in the magnitude of their responses, as well as in key competitive attributes such as relative growth rate (RGR, Grime & Hunt 1975). These differences may help to explain some of the responses we observed. For instance, the positive effects of species evenness observed when *Lolium* was dominant in the 4:1:1 mixtures may be partially explained by the differences in RGR between species (*Lolium* has a lower RGR than *Holcus* and *Plantago*, Grime & Hunt, 1975). It has been shown that productivity responses of assemblages to global change drivers are strongly dependent on species composition (He et al. 2002; Craine et al. 2003). Our results also point in this direction, suggesting that species composition is a relevant factor to fully disentangle the multiplicity of potential responses of plant assemblages to global change.

Studies evaluating the effects of species evenness on the productivity of grassland assemblages have found that these range from negative to positive, and that they are largely dependent upon species-specific traits (Wilsey & Potvin 2000; Polley et al. 2003; Mulder et al. 2004). Our results agree with these findings, as the magnitude and direction of the observed effects of species evenness on productivity (either negative, neutral or positive) were dependent on the identity of the dominant species in the mixtures (Table 2). It has been suggested that positive effects of species evenness are mediated by greater complementarity in resource use (Mulder et al. 2004). Albeit we found higher nitrogen concentration under maximum evenness in the *Holcus*-dominated mixtures, our results link the observed responses with patterns of biomass allocation. When *Lolium* was the dominant species in mixtures –the only situation where increasing evenness led to an increase in biomass–, the assemblages allocated less biomass to roots with increasing species evenness. Such a response should have reduced belowground competition (Aerts et al. 1991), and this amelioration could have promoted the increase in aboveground biomass observed in the maximum evenness mixtures.

The effects of water heterogeneity on productivity found in the Experiment 4 (from positive to negative as water availability increased) parallel those of Lundholm and Larson (2004), who found that effects of water heterogeneity on the productivity of herbaceous assemblages varied from neutral to negative as water availability increased. These results contrast with studies carried out in mesic grasslands of North America, where Fay et al. (2003), in a rainfall manipulation experiment, found that increasing variability in rainfall negatively affected productivity at either ambient or reduced (70% of ambient) rainfall levels. A recent analysis by Nippert et al. (2006) of long-term data gathered in the same system showed a positive and a negative effect of rainfall variability early and later in the growing season, respectively. These authors suggest that these seasonal patterns may be driven by differences in overall water availability, with increases in variability having a negative effect under low water availability levels. Discrepancies may be due to differences in the duration, experimental approach and species involved. Despite them, our results (and those from previous studies) highlight the importance of temporal rainfall patterns, not just overall means, on grassland productivity. Further, we note that the magnitude of the response to water heterogeneity under low water availability levels was modulated by nutrient heterogeneity. To our knowledge, this is the first time that such an interaction has been reported.

Responses at the Species Level

At the species level, the magnitude of the responses to each of the factors evaluated was species-specific, suggesting that they affected the interspecific competitive relationships between the species evaluated (Fransen et al. 2001; Novoplansky and Goldberg 2001). Nutrient heterogeneity was a key factor determining the composition of the assemblages, as it affected the contribution to aboveground biomass in four of the five species forming the assemblages in Experiment 4, and interacted with nutrient availability, water heterogeneity and [CO_2] to determine these responses.

In the Experiment 1, *Holcus* was the species that responded most to elevated CO_2 under homogeneous conditions at the medium and highest nutrient availability conditions. However, under heterogeneous conditions of nutrient supply the subordinate species responded most to elevated [CO_2] (Figure 6). In a review of competition experiments, Poorter & Navas (2003) found that C3 grasses and N_2-fixing species were the groups of species that benefited most from elevated [CO_2] under high and low nutrient supply, respectively.

Interestingly, and contrary to this general pattern, we found that the N_2-fixer *Trifolium* benefited more from elevated [CO_2] than the whole assemblage under high nutrient availability conditions, but only when the nutrients were supplied heterogeneously. Previous studies have found that *Trifolium* responds to both increases in soil nutrient availability and heterogeneity by increasing biomass growth (Turkington et al. 1991; Hutchings et al. 1997) but the joint effects of both variables on this species when growing in competition has not been evaluated before. Our results are consistent with these studies, and suggest that a plastic response of *Trifolium* to the joint presence of increased nutrient availability and heterogeneity may be non-additive under elevated [CO_2]. As legumes can transfer the nitrogen fixed from the atmosphere to other species, they are of paramount importance as drivers of community-wide responses to elevated CO_2 (Navas et al. 1997; Warwick et al. 1998; Stöcklin & Körner 1999). Thus, the evaluation of the role of soil heterogeneity as a driver of legume responses to joint increases in the atmospheric concentration of [CO_2] and nutrient availability merits further study.

Consistent with the findings of Berntson et al. (1998) for a model annual community, the enhancement of nitrogen uptake was linearly related to biomass production. This suggests that the ability of plants to increase nitrogen uptake may be an important determinant of which species in an assemblage will be able to respond to elevated [CO_2] with increased biomass production. Notably, the greatest enhancement of both biomass and nutrient uptake under elevated [CO_2] was found for the subordinate species under conditions of high nutrient availability and heterogeneous nutrient supply. Others (e.g., Navas et al. 1997; Berntson et al. 1998; Stöcklin & Körner 1999) have also found that subordinate species are the most responsive to an increase in [CO_2]. Our results add to this literature by emphasizing that such responses may be dependent on both soil nutrient heterogeneity and availability. If the patterns observed in our experiment hold true, species composition is likely to change under elevated [CO_2], in a manner that is dependent on nutrient availability and spatial pattern. Changes in species composition under elevated [CO_2] have been detected in both field and microcosm studies (Berntson et al. 1998; Jongen & Jones 1998; He et al. 2002), but none of these studies have explicitly considered soil heterogeneity as a factor in the outcome of these changes.

Extrapolation of the Results Obtained to Field Conditions

In the field, heterogeneity in resource distribution arises as a result of organic inputs and their subsequent microbial decomposition. Such decomposition will release nutrients for plant capture in a spatio-temporal fashion more complex than that resulting from placing a patch of inorganic nutrients directly in the soil, the most utilized approach in experimental work on soil nutrient heterogeneity (Hutchings et al. 2000b). Given that we used natural soil and organic material, the decomposition of the organic material in our experiment may most closely resemble that observed in the field (Hodge 2004). However, our approach is not without limitations. For example: (a) growing assemblages in PVC piping may alter root foraging responses due to physical restriction of lateral root growth (see Fransen et al. 1999); (b) patterns of patch heterogeneity and degree of contrast to the background soil may not reflect those found in the field; and (c) standardised climatic conditions (i.e. light, temperature, humidity, water supply) may amplify plant responses to soil nutrient availability and heterogeneity over those that may be observed in the field. The later limitation is applicable to many studies using CEFs and, when taken together with the first two limitations, it is clear that extrapolation of our results to the natural world should be done with caution. Nevertheless, we are confident that our findings demonstrate potential plant responses to simultaneous changes in soil heterogeneity and global change drivers (Jones et al. 2000).

It is interesting to note that some of the main effects and interactions observed were experiment-specific (Tables 1 and 2). For instance, responses of the below: aboveground biomass ratio to soil heterogeneity and [CO_2] varied from negative to neutral. The experiments differed among them in important aspects, including total plant density, number of species forming the assemblages, nutrient availability levels, and growing conditions (characteristics of the organic material used as fertilizer, light intensity and watering regime). Given the strong influence of both environmental context and plant-plant interactions on plant responses to nutrient heterogeneity (Wijeshinghe et al. 2001) and global change drivers (Poorter & Navas 2003), these discrepancies are not fully surprising. Further research is thus needed to evaluate the generality of the results presented here.

Conclusion

Our working hypothesis, i.e., that that soil heterogeneity modulates the magnitude of the response of both individuals and assemblages to multiple global change drivers, was supported by our results. In all the experiments conducted we did find a series of two- and three-term interactions determining key responses at both the individual and assemblage level. These results strongly suggest that single-factor studies should not be used to extrapolate how grassland ecosystems may respond to global change (see also Körner 2003 and Kardol 2012), which is composed of multiple—and potentially interacting—drivers.

A substantial part of the research conducted to date to evaluate the effects of global change on ecosystem processes in terrestrial ecosystems has used grassland species (e.g. Loreau et al. 2002; Dukes et al. 2005; He et al. 2002; Reich et al. 2004; Bradford et al. 2012), which are known to be highly responsive to soil heterogeneity (Robinson & Van Vuuren 1998). Interestingly, the experiments presented here are among the first to consider the effects of soil heterogeneity on the observed responses of grasslands to global change drivers (see García-Palacios et al. 2012 for a review). Given the high economic costs and logistic constrains associated to conducting experiments such as presented here in the field, microcosm experiments conducted in CEFs will continue playing a pivotal role in future studies on the topic. These studies should: (i) evaluate the generality of our findings, (ii) test if similar responses are observed with species belonging to other functional and community types, and (iii) expand the suite of response variables measured to include key processes not considered in our research, such as root dynamics, microbial processes and nutrient cycling (see García-Palacios et al. 2011, 2013 for examples). Such studies will undoubtedly increase our ability to predict how species and ecosystems will respond to the ongoing global change in the real, and highly heterogeneous, world.

ACKNOWLEDGMENTS

We thank María D. Puche, Mark Bradford, José J. Maestre, Anne Rosenbarger, Lea Harrell, José L. Quero, Rob Jackson, Paul Heine, Heather Hemrick, Richard Mackubin, David Tremmel and the Duke Phytotron staff (Jerome Smith, Todd Smith, Norman Hill and Marcia Kirinius) for their help during the different phases of the experiments summarized in this chapter. FTM was supported by a Fulbright fellowship (FU2003-0398) and a Ramón y Cajal contract from the Spanish Ministerio de Educación y Ciencia during the realization of the experiments and the writing of this chapter, respectively. The research underlying this chapter was supported in part by USDA Specific Co-operative Agreement #58-1270-3-070, and by NSF-DEB-02-12123, NSF-IBN-99-85877 to the Duke Phytotron and NSF-SBR-9521914 (Subcontract # 538819-55801 from Carnegie Mellon University) grants. FTM acknowledges support from the European Research Council under the European Community's Seventh Framework Programme (FP7/2007-2013)/ERC Grant agreement no. 242658 (BIOCOM).

REFERENCES

Aerts, R., Boot, R. G. A. & Vanderaart, P. J. M. (1991) The relation between aboveground and belowground biomass allocation patterns and competitive ability. *Oecologia*, 87, 551-559.

Armas, C., Pugnaire, F. I. & Ordiales, R. (2004) Measuring plant interactions: A new comparative index. *Ecology*, 85, 2682-2686.

Arnone, J.A. III (1997) Temporal responses of community fine root populations to long-term elevated atmospheric CO_2 and soil nutrient patches in model tropical ecosystems. *Acta Oecologica,* 18, 367-376.

Bassirirad, H., Gutschick, V.P. & Lussenhop, J. (2001) Root system adjustments: Regulation of plant nutrient uptake and growth responses to elevated CO_2. *Oecologia*, 126, 305-320.

Bazzaz, F.A. (1990) The response of natural ecosystems to the rising global CO_2 levels. *Annual Review of Ecology and Systematics*, 21, 167-196.

Berntson, G.M., Rajakaruna, N. & Bazzaz, F.A. (1998) Growth and nitrogen uptake in an experimental community of annuals exposed to elevated atmospheric CO_2. *Global Change Biology*, 4, 607-626.

Bliss, K.M., Jones, R.H., Mitchell, R. & Mou, P. (2002) Are competitive interactions influenced by spatial nutrient heterogeneity and root foraging behavior? *New Phytologist*, 154, 409-417.

Bradford, M. A., Wood, S., Maestre, F. T., Reynolds, J. F. & Warren III, R. (2012) Contingency in ecosystem but not plant community response to multiple global change factors. *New Phytologist*, 196, 462-471.

Cain, M.L., Subler, S., Evans, J.P. & Fortin, M.J. (1999) Sampling spatial and temporal variation in soil nitrogen availability. *Oecologia*, 118, 397-404.

Casper, B.B. & Cahill, J.F. Jr (1996) Limited effects of soil nutrient heterogeneity on populations of *Abutilon theophrasti* (Malvaceae). *American Journal of Botany*, 83, 333-341.

Casper, B.B. & Cahill, J.F. Jr. (1998) Population-level responses to nutrient heterogeneity and density by *Abutilon theophrasti* (Malvaceae): An experimental neighborhood approach. *American Journal of Botany*, 85, 1680-1687.

Clark, H., Newton, P.C.D. & Barker, D.J. (1999) Physiological and morphological responses to elevated CO_2 and a soil moisture deficit of temperate pasture species growing in an established plant community. *Journal of Experimental Botany*, 50, 233-242.

Cotrufo, M.F., Ineson, P. & Scott, A. (1998) Elevated CO_2 reduces the nitrogen concentration of plant tissues. Global Change Biology, 4, 43-54.

Craine, J.M., Reich, P.B., Tilman, D., Ellsworth, D.S., Fargione, J., Knops, J. & Naeem, S. (2003) The role of plant species in biomass production and response to elevated CO_2 and N. *Ecology Letters*, 6, 623-630.

Day, K.J., Hutchings, M.J. & John, E.A. (2003) The effects of spatial pattern of nutrient supply on yield, structure and mortality in plant populations. *Journal of Ecology*, 91, 541-553.

Désilets, P. & Houle, G. (2005) Effects of resource availability and heterogeneity on the slope of the species-area curve along a floodplain-upland gradient. *Journal of Vegetation Science*, 16, 487-496.

Dukes, J.S., Chiariello, N.R., Cleland, E.E., Moore, L.A., Shaw, M.R., Thayer, S., Tobeck, T., Mooney, H.A. & Field, C.B. (2005) Responses of grassland production to single and multiple global environmental changes. *PLoS Biology,* 3, e319.

Easterling, D.R., Meehl, G.A., Parmesan, C., Changnon, S.A., Karl, T.R. & Mearns, L.O. (2000) Climate extremes: observation, modeling, and impact. *Science*, 289, 2068-2074.

Einsmann, J. C., Jones, R. H., Pu, M. & Mitchell, R. J. (1999) Nutrient foraging traits in 10 co-occurring plant species of contrasting life forms. *Journal of Ecology*, 87, 609-619.

Fay, P. A., Carlisle, J. D., Knapp, A. K., Blair, J. M. & Collins, S. L. (2003) Productivity responses to altered rainfall patterns in a C_4-dominated grassland. *Oecologia*, 137, 245-251.

Farley, R.A. & Fitter, A.H. (1999) Temporal and spatial variation in soil resources in a deciduous woodland. *Journal of Ecology*, 87, 688-696.

Fowler, N. & Antonovics, J. (1981) Competition and coexistence in a North Carolina grassland. I. Patterns in undisturbed vegetation. *Journal of Ecology*, 69, 825-841.

Fransen, B., De Kroon, H., De Kovel. C.G. & Van Den Bosch, F. (1999) Disentangling the effects of root foraging and inherent growth rate on plant biomass accumulation in heterogeneous environments: A modelling study. *Annals of Botany*, 84, 305-311.

Fransen, B., De Kroon, H. & Berendse, F. (2001) Soil nutrient heterogeneity alters competition between two perennial grass species. *Ecology*, 82, 2534-2546.

García-Palacios, P., Maestre, F.T., Bardgett, R. & De Kroon, H. (2012) Plant responses to soil heterogeneity and global environmental change. *Journal of Ecology*, 100, 1303–1314.

García-Palacios, P., Maestre, F. T. & Gallardo, A. (2011) Soil nutrient heterogeneity modulates ecosystem responses to changes in the identity and richness of plant functional groups. *Journal of Ecology*, 99, 551–562.

García-Palacios, P., Maestre, F. T. & Milla, R. (2013) Community-aggregated plant traits interact with soil nutrient heterogeneity to determine ecosystem functioning. *Plant and Soil*, 364, 119–129.

Grime, J.P. (1979) Plant strategies and vegetation processes. Wiley & Sons, Chichester

Grime, J.P. & Hunt, R. (1975) Relative growth rate: its range and adaptative significance in a local flora. *Journal of Ecology*, 63, 393-422.

Grime, J.P. & Mackey, J.M.L. (2002) The role of plasticity in resource capture by plants. *Evolutionary Ecology*, 16, 299-307.

Grünzweig, J.M. & Körner, C.h. (2001) Growth, water and nitrogen relations in grassland model ecosystems of the semi-arid Negev of Israel exposed to elevated CO_2. *Oecologia*, 128, 251-262.

Grünzweig, J.M. & Körner, C.h. (2003) Differential phosphorus and nitrogen effects drive species and community responses to elevated CO_2 in semi-arid grassland. *Functional Ecology*, 17, 766-777.

Hanson, P.J., Wullschleger, S.D., Norby, R.J., Tschaplinski, T.J. & Gunderson, C.A. (2005) Importance of changing CO_2, temperature, precipitation, and ozone on carbon and water cycles of an upland-oak forest: incorporating experimental results into model simulations. *Global Change Biology*, 11, 1402-1423.

He, J.S., Bazzaz, F.A. & Schmid, B. (2002) Interactive effects of diversity, nutrients and elevated CO_2 on experimental plant communities. *Oikos*, 97, 337-348.

Hodge, A. (2004) The plastic plant: root responses to heterogeneous supplies of nutrients. *New Phytologist*, 162, 9-24.

Hodge, A., Stewart, J., Robinson, D., Griffiths, B.S. & Fitter, A.H. (2000) Competition between roots and soil micro-organisms for nutrients from nitrogen-rich patches of varying complexity. *Journal of Ecology*, 88, 150-164.

Hooper, D.U., Chapin, F.S. III, Ewel, J.J., Hector, A., Inchausti, P., Lavorel, S., Lawton, J.H., Lodge, D.M., Loreau, M., Naeem, S., Schmid, B., Setäla, H., Symstad, A.J., Vandermeer, J. & Wardle, D.A. (2005) Effects of biodiversity on ecosystem functioning: A consensus of current knowledge. *Ecological Monographs*, 75, 3-35.

Huber-Sannwald, E. & Jackson, R.B. (2001) Heterogeneous soil-resource distribution and plant responses-from individual-plant growth to ecosystem functioning. *Progress in Botany*, 62, 451-476.

Hutchings, M.J., John, E.A. & Stewart, A.J.A. (2000a) The ecological consequences of environmental heterogeneity. Blackwell Science, Oxford

Hutchings, M.J., John, E.A. & Wijesinghe, D.K. (2003) Toward understanding the consequences of soil heterogeneity for plant populations and communities. *Ecology*, 84, 2322-2334.

Hutchings, M.J., Turkington, R., Carey, P. & Klein, E. (1997) Morphological plasticity in Trifolium repens L.: The effects of clone genotype, soil nutrient level, and the genotype of conspecific neighbours. Canadian *Journal of Botany*, 75, 1382-1393.

Hutchings, M.J., Wijesinghe, D.K. & John, E.A. (2000b) The effects of heterogeneous nutrient supply on plant performance: a survey of responses, with special reference to clonal herbs. In: Hutchings, M.J., John, E.A. & Stewart, A.J.A. (eds) The ecological consequences of environmental heterogeneity. Blackwell Science, Oxford, pp. 91-110.

Jackson, R.B., & Caldwell, M.M. (1993) Geostatistical patterns of soil heterogeneity around individual perennial plants. *Journal of Ecology*, 81, 683-692.

Jones, T.H., Bezemer, T.M., Körner, Ch., Lawton, J.H. & Thompson, L.J. (2000) Comparing studies of artificial and natural ecosystem responses to CO_2 enrichments. *Biotronics*, 29, 1-7.

Jongen, M. & Jones, M.B. (1998) Effects of elevated carbon dioxide on plant biomass production and competition in a simulated neutral grassland community. *Annals of Botany*, 82, 111-123.

Joshi, J., Matthies, D. & Schmid, B. (2000) Root hemiparasites and plant diversity in experimental grassland communities. *Journal of Ecology*, 88, 634-644.

Kardol, P., De Long, J. R. & Sundqvist, M. K. (2012) Crossing the threshold: the power of multi-level experiments in identifying global change responses. *New Phytologist*, 196, 323-326.

Kembel, S.W. & Cahill, J.F. Jr. (2005) Plant phenotypic plasticity belowground: A phylogenetic perspective on root foraging trade-offs. *American Naturalist*, 166, 216-230.

Kirkham, F.W., Mountford, J.O, & Wilkins, R.J. (1996) The effects of nitrogen, potassium and phosphorus addition on the vegetation of a Somerset peat moor under cutting management. *Journal of Applied Ecology*, 33, 1013-1029.

Körner, Ch. (2001) Experimental plant ecology: some lessons from global change research. In: Ecology: Achievement and Challenge (eds Press MC, Huntly NJ, Levin S), pp. 227-248. Blackwell Science, Oxford, UK

Körner, Ch. (2003) Nutrients and sink activity drive plant CO_2 responses: Caution with literature-based analysis. *New Phytologist*, 159, 537-538.

Körner, Ch., Bazzaz, F.A. (eds) (1996) Carbon Dioxide, Population, and Community. Academic Press, San Diego

Lawton, J.H. (1995) Ecological experiments with model systems. *Science*, 269, 328-331.

Loreau, M., Naeem, S. & Inchausti, P. (eds) (2002) Biodiversity and Ecosystem Functioning: Synthesis and Perspectives. Oxford University Press, Oxford

Lundholm, J.T. & Larson, D.W. (2004) Experimental separation of resource quantity from temporal variability: seedling responses to water pulses. *Oecologia*, 141, 346-352.

Luo, Y., Reynolds, J.F., Wang, Y. & Wolfe, D. (1999) A search for predictive understanding of plant responses to elevated [CO_2]. *Global Change Biology*, 5, 143-156.

Maestre, F.T., Bradford, M.A. & Reynolds, J.F. (2005) Soil nutrient heterogeneity interacts with elevated CO_2 and nutrient availability to determine species and assemblage responses in a model grassland community. *New Phytologist*, 168, 637-650.

Maestre, F.T., Bradford, M.A. & Reynolds, J.F. (2006) Soil heterogeneity and community composition jointly influence grassland biomass. *Journal of Vegetation Science*, 17, 261-270.

Maestre, F.T., Cortina, J., Bautista, S., Bellot, J., Vallejo, R. (2003) Small-scale environmental heterogeneity and spatio-temporal dynamics of seedling survival in a degraded semiarid ecosystem. *Ecosystems*, 6, 630-643.

Maestre, F.T., Quero, J.L., Valladares, F. & Reynolds, J.F. (2007) Individual vs. population plastic responses to elevated CO_2, nutrient availability and heterogeneity: a microcosm experiment with co-occurring species. *Plant and Soil*, 296, 53-64.

Maestre, F.T. & Reynolds, J.F. (2006a) Nutrient availability and atmospheric CO_2 partial pressure modulate the effects of nutrient heterogeneity on the size structure of populations in grassland species. *Annals of Botany*, 98, 227-235.

Maestre, F.T. & Reynolds, J.F. (2006b) Spatial heterogeneity in nutrient supply modulates plant nutrient and biomass responses to multiple global change drivers in model grassland communities. *Global Change Biology*, 12, 2431-2441.

Maestre, F.T. & Reynolds, J.F. (2007a) Biomass responses to elevated CO_2, soil heterogeneity and diversity: an experimental assessment with grassland assemblages. *Oecologia*, 151, 512-520.

Maestre, F.T. & Reynolds, J.F. (2007b) Amount or pattern? Grassland responses to the heterogeneity and availability of two key resources. *Ecology*, 88, 501-511.

Mulder, C.P.H., Bazeley-White, E., Dimitrakopoulos, P.G., Hector, A., Scherer-Lorenzen, M. & Schmid, B. (2004) Species evenness and productivity in experimental plant communities. *Oikos*, 107, 50-63.

Navas, M.L., Sonie, L., Richarte, J. & Roy, J. (1997) The influence of elevated CO_2 on species phenology, growth and reproduction in a Mediterranean old-field community. *Global Change Biol*, 3, 523-530.

Niklaus, P.A., Körner, Ch. (2004) Synthesis of a six-year study of calcareous grassland responses to in situ CO_2 enrichment. *Ecological Monographs*, 74, 491-511.

Nippert, J.B., Knapp, A.K., Briggs, J.M. (2006) Intra-annual rainfall variability and grassland productivity: can the past predict the future? *Plant Ecology*, 184, 65-74,

Norby, R.J. & Luo, Y. (2004) Evaluating ecosystem responses to rising atmospheric CO_2 and global warming in a multi-factor world. *New Phytologist*, 162, 281-293.

Novoplansky, A. & Goldberg, D.E. (2001) Effects of water pulsing on individual performance and competitive hierarchies in plants. *Journal of Vegetation Science*, 12, 199-208.

Pan, D., Bouchard, A., Legendre, P. & Domon, G. (1998) Influence of edaphic factors on the spatial structure of inland halophytic communities: A case study in China. *Journal of Vegetation Science*, 9, 797-804.

Polley, H.W., Wilsey, B.J. & Derner, J.D. (2003) Do species evenness and plant density influence the magnitude of selection and complementarity effects in annual plant species mixtures?. *Ecology Letters*, 6, 248-256.

Poorter, H. (1993) Interspecific variation in the growth response of plants to an elevated ambient CO_2 concentration. *Vegetatio*, 104/105, 77-97.

Poorter, H. & Nagel, O. (2000) The role of biomass allocation in the growth response of plants to different levels of light, CO_2, nutrients and water: a quantitative review. *Australian Journal of Plant Physiology*, 27, 595-607.

Poorter, H. & Navas, M.L. (2003) Plant growth and competition at elevated CO_2: On winners, losers and functional groups. *New Phytologist*, 157, 175-198.

Rajaniemi, T. & Reynolds, H.L. (2004) Root foraging for patchy resources in eight herbaceous plant species. *Oecologia*, 141, 514-525.

Reich, P.B., Knops, J., Tilman, D., Craine, J., Ellsworth, D., Tjoelker, M., Lee, T., Wedin, D., Naeem, S., Bahauddin, D., Hendrey, G., Jose, S., Wrage, K., Goth, J. & Bengston, W. (2001) Plant diversity enhances ecosystem responses to elevated CO_2 and nitrogen deposition. *Nature*, 410, 809-812.

Reich, P.B., Tilman, D., Naeem, S., Ellsworth, D.S., Knops, J., Craine, J., Wedin, D. & Trost, J. (2004) Species and functional group diversity independently influence biomass accumulation and its response to CO_2 and N. *Proceedings of the National Academy of Sciences,* USA 101, 10101-10106.

Robinson, D. (1994) The responses of plants to non-uniform supplies of nutrients. New Phytologist, 127, 635-674.

Robinson, D., Hodge, A., Griffiths, B.S. & Fitter, A.H. (1999) Plant root proliferation in nitrogen-rich patches confers competitive advantage. *Proceedings of the Royal Society of London B*, 266, 431-435.

Robinson, D. & Van Vuuren, M.M.I. (1998) Responses of wild plants to nutrient patches in relation to growth rate and life-form. In: Lambers, H., Poorter, H., van Vuuren, M.M.I. (eds) Variation in plant growth. Backhuys, Leiden, pp. 237-257.

Ryel, R.J., Caldwell, M.M. & Manwaring, J.H. (1996) Temporal dynamics of soil spatial heterogeneity in sagebrush-wheatgrass steppe during a growing season. *Plant and Soil*, 184 299-306.

Sala, O.E., Chapin, F.S. III, Armesto, J.J. et al. (2000) Global biodiversity scenarios for the year 2100. *Science*, 287, 1770-1774.

Shaw, M.R., Zavaleta, E.S., Chiariello, N.R., Cleland, E.E., Mooney, H.A. & Field, C.B. (2002) Grassland responses to global environmental changes suppressed by elevated CO_2. *Science*, 298, 1987-1990.

Stein, A., Brouwer, J. & Bouma, J. (1997) Methods for comparing spatial variability patterns of millet yield and soil data. *Soil Science Society of America Journal*, 61, 861-870.

Stöcklin, J., Schweizer, K. & Körner, Ch. (1998) Effects of elevated CO_2 and phosphorus addition on productivity and community composition of intact monoliths from calcareous grasslands. *Oecologia*, 116, 50-56.

Stöcklin, J. & Körner, Ch. (1999) Interactive effects of elevated CO_2, P availability and legume presence on calcareous grassland: results of a glasshouse experiment. *Functional Ecology*, 13, 200-209.

Strain, B.R. & Cure, J.D. (1994) Direct effects of atmospheric CO_2 enrichment on plants and ecosystems. An updated bibliographic data base. Oak Ridge National Laboratory, Oak Ridge, Tennesse

Tilman, D. (1988). Plant Strategies and the Dynamics and Structure of Plant Communities. Princenton University Press, Princenton

Turkington, R., Sackville Hamilton, R. & Gliddon, C. (1991) Whitin-population variation in localized and integrated responses of *Trifolium repens* to biotically patchy environments. *Oecologia*, 86, 183-192.

Vitousek, P.M., Mooney, H.A., Lubchencko, S. & Melillo, J.M. (1997) Human domination of the Earth's ecosystems. *Science*, 277, 494–499.

Warwick, K.R., Taylor, G. & Blum, H. (1998) Biomass and compositional changes occur in chalk grassland turves exposed to elevated CO_2 for two seasons in FACE. *Global Change Biology*, 4, 375-385.

Wijesinghe, D.K., John, E.A., Beruskens. S. & Hutchings, M.J. (2001) Root system size and precision in nutrient foraging: responses to spatial pattern of nutrient supply in six herbaceous species. *Journal of Ecology*, 89, 972-983.

Wijesinghe, D.K., John, E.A. & Hutchings, M.J. (2005) Does pattern of soil resource heterogeneity determine plant community structure? An experimental investigation. *Journal of Ecology*, 93, 99-112.

Wilsey, B.J. & Potvin, C. (2000) Biodiversity and ecosystem functioning: Importance of species evenness in an old field. *Ecology*, 81, 887-892.

Wurst, S., Langel, R., Reineking, A., Bonkowski, M., Scheu, S. (2003) Effects of earthworms and organic litter distribution on plant performance and aphid reproduction. *Oecologia*, 137, 90-96.

Zavaleta, E.S., Shaw, M.R., Chiariello, N.R., Mooney, H.A., Field, C.B. (2003) Additive effects of simulated climate changes, elevated CO_2, and nitrogen deposition on grassland diversity. *Proceedings of the National Academy of Sciences* USA, 100, 7650-7654.

In: Microcosms
Editor: Christopher C. Harris
ISBN: 978-1-62618-661-3

Chapter 3

DEVELOPMENT AND OPTIMISATION OF AN AQUATIC LABORATORY MICROCOSM FOR ECOTOXICOLOGICAL RISK ASSESSMENT

Bernard Clément[1], Gaëlle Triffault-Bouchet[2] and Hélène Delhaye[1]
[1]Université de Lyon, UMR5023 Ecologie des Hydrosystèmes Naturels et Anthropisés, Université Lyon 1, ENTPE, CNRS, Vaulx en Velin, France
[2]Centre d'expertise en analyse environnementale du Québec, Ministère du Développement durable, de l'environnement et des Parcs, Division Écotoxicologie et Évaluation du risque, Sainte Foy, Québec, Canada

ABSTRACT

Ecotoxicological assessment of chemical substances is mainly based on standard bioassays, most often single-species tests covering the range of acute and chronic toxicity. In our laboratory we have been developping since 1997 a protocol of ecotoxicological bioassay in 2-L laboratory microcosms and applied it to the study of various pollutants and scenarios of ecotoxicological risk assessment in the field of urban facilities and transport infrastructures. Effects are assessed on five different organisms (micro-algae, duckweeds, daphnids, amphipods, chironomids) using endpoints such as growth, emergence (chironomids), reproduction (daphnids), survival, with a duration exposure of 3-4 weeks. For specific studies, some variants of the 2-L protocol have been developed in higher volumes (up to 100 L) with increasing biological diversity and different hydraulic conditions. More recently, a flow-through microcosm assay (dynamic assay) was developped for the 2-L protocol to improve conditions inside the microcosms, resulting in stabilisation of physico-chemical parameters, increase of organisms fitness and reduction of variability.

This chapter will present the various protocols, their application to different ecotoxicological risk assessment studies, their limitations (variability and sensitivity, ecological relevance, organism fitness, ...), and the improvement of the 2-L protocol through continuous water renewal. Finally, recent works and perspectives in modelling will be presented.

1. INTRODUCTION

One of the main objectives of ecotoxicology is to predict the effects of toxic substances on receiving ecosystems. This requires tools and methods whose pertinence, depending on the objective, is inversely proportional to their cost and simplicity. The simplest tools to implement are single-species tests carried out in the laboratory under standardized conditions. These consist of studying a single species considered as representative of a class of organisms of a given trophic level and/or as a privileged target of the toxic substance due to its place or role in the ecosystem. These tests are characterized by their simplicity and speed of implementation, their sensitivity, their repeatability and their moderate cost (Kimball and Levin, 1985; Calow, 1993). Biotic (species, age) and abiotic (temperature, photoperiod, type of lighting, exposure time, physico-chemical composition of the environment) conditions are controlled and standardized, which enables comparing the results obtained, particularly with different pollutants and between different laboratories (Taub, 1989a). Even when used in a battery of bioassays, they cannot be used as tools to predict the effects of a toxic substance on natural ecosystems (Taub, 1997). The response of the latter cannot be predicted on the basis of the response of several species tested separately in the laboratory, since the way natural ecosystems function depends on the processes involved at different levels of biological organization, those of individual organisms, populations, communities and the ecosystem itself (Cairns et al., 1992). Therefore, taken alone, they do not represent a complete enough approach to predict risks for the environment, but nonetheless remain essential (Pratt et al., 1987).

Contrary to single-species tests, *in situ* studies enable studying the impact of the contamination of an environment by one or more toxic substances under real conditions and at full-scale, but they generate enormous methodological problems (Barry and Logan, 1998) and are very costly.

Experimental systems (microcosms and mesocosms) in which one or several variables can be controlled prove to be very useful for assessments intended for prediction. Tests in mesocosms and outdoor microcosms (reconstructed ecosystems with volumes varying from a few m^3 to several dozens or hundreds of m^3 and generally subject to climatic hazards and to natural colonization (Larsen et al., 1986; Lay et al., 1993; Caquet et al., 1996; Giddings et al., 1994; Heimbach et al., 1992; Tidou et al., 1992)) enable partially solving the methodological problems caused by *in situ* studies (controlled contamination, possibility of reference systems, etc.). Tests in laboratory microcosms (a few liters to several dozen liters placed under controlled temperature and lighting conditions), also called multi-species tests when they implement several species, constitute a good compromise in terms of cost and ecotoxicological pertinence between single-species tests and outdoor microcosm tests (Barry and Logan, 1998). Indeed, they offer a large number of effect criteria and take into account interactions between species (competition, predation, etc.) while presenting a sufficient level of standardization and replicability (Cairns and Cherry, 1993).

There are relatively few methods in the literature on laboratory microcosms that provide a schematic simulation of a stagnant water ecosystem at different trophic levels and which present the qualities required from any ecotoxicological test (convenience of implementation, reproducibility, sensitivity, etc.). The studies performed at Savoie University on microcosms of a few liters over several years to examine the fate and ecotoxic effects of substances and effluents (Merlin and Blake, 1991; Merlin et al., 1992, 1993, 1994; Yosowidagdo, 1996) have provided a good starting point. These microcosms were most often composed of a water column (synthetic environment), in which microalgae, duckweed and cladocerans (daphnia) developed, and sediment (Fontainebleau sand + chitin) which constituted a medium for established macrophytes (bryophyte, *Elodea*, *Groenlandia*), mollusks (gastropods) and crustaceans (*Asellus aquaticus*). Mention can also be made of the Standardized Aquatic Microcosms (SAM) (Conquest and Taub, 1989; Taub, 1989a, 1989b), which are 3-liter systems for testing the effects of toxic substances on a diversified community of laboratory organisms for 63 days (10 species of algae, daphnia, amphipods, ostracods, protozoans, rotifers). Despite it being a standardized method, this protocol is cumbersome to implement, requiring good mastery of cultures and breeding, as well as a great deal of time and energy to monitor the systems and collect the data. Moreover, like the sediment components used with the microcosms developed at Savoie University, that used in this method is reduced to a physical medium (sand + chitin + cellulose) for certain epibenthic organisms (algae, amphipods and ostracods) but does not contain any benthic organisms (i.e. living inside the sediment).

The aim of this chapter, after providing a short bibliographical overview of microcosm tests, is to present the results obtained in this field by the authors, by underlining the lessons learned from previous works and the improvements made to the basic protocol. It will conclude on the perspectives given by numerical modelling, completing the experimental model represented by the microcosm. Most of the authors are at the Laboratory of Environmental Sciences which recently became the “Effects of Pollutants on Ecosystems” team of the Laboratory Ecology of Natural and Man-impacted Hydrosystems (LEHNA, Université Lyon 1, ISARA and ENTPE, UMR CNRS 5029).

2. Aquatic Microcosms in Ecotoxicology

2.1. The Place of Microcosm Tests in the Literature on Ecotoxicology

The works published on the theme “laboratory microcosm simulating lentic freshwater ecosystems” represent only 6% of the works on freshwater ecotoxicology (Clément, 2006). By the term microcosm, we refer to a reconstructed ecosystem in which one or more species are exposed simultaneously to one or more toxic substances and interact with each other and with the environment. For freshwater environments, numerous works refer to microcosm and outdoor mesocosm tests, but this large number should be viewed in relative terms given the complexity and the duration of the experiments. Consequently, whether regarding laboratory microcosm and outdoor micro- or mesocosm tests, these works represent a small proportion of the experiments conducted in the field of aquatic ecotoxicology, as the vast majority of them focus on lower levels of complexity (individual, cell) exposed under simplified and often standardized conditions.

2.2. Objectives of the Microcosm Tests in the Literature on Ecotoxicology

In most cases, a microcosm is used to study the direct and indirect effects of one or more pollutants on several species exposed simultaneously, belonging to different levels of trophic organization and interacting with each other and with the environment, and/or with the major biological functions characterizing an ecosystem : respiration, photosynthesis, primary productivity, mineralization of organic matter. The fact that populations are exposed *simultaneously* has several advantages. Compared to a battery of single-species tests that would be based on the species of the microcosm, the populations evolving in the same compartment (water column or sediment) are exposed to identical conditions (oxygen content and redox potential, pH, physico-chemical characteristics of the water and the sediment, lighting, temperature, etc.) and to the same chemical species of the pollutants introduced, which is not generally the case with a battery of tests, because of the protocols specific to each test (starting with the testing environment). For two interacting species (e.g., prey-predator relation), simultaneous exposure enables not only taking into account the effect of the pollutant on each species (direct effects) but also on the relation between the two species, thereby making it possible to evaluate the indirect effects of the pollutant on an ecological factor and on the dynamics of the populations. A toxic substance may not have a direct effect on a tolerant species but may affect it indirectly through certain ecological mechanisms (reduced predation pressure, increased or decreased sensitivity to predation, Fleeger et al., 2003).

Working with several species belonging to the same trophic level enables highlighting effects on dominance or diversity. When two phytoplanktonivorous species are present, partial mortality of one of them can lead to its extinction (it has a competitive disadvantage), whereas that same species would remain if it were alone. The microcosm can be used to reproduce certain functional activities of ecosystems (primary production, mineralization of organic matter), thus it enables evaluating the effects of toxic substances on these activities. It is also possible to take into account the effect of certain biotic activities on the fate of the substance tested. Madsen et al. (1997) studied the impact of fluoranthene on a mixture by bioadvection of sediment particles by the sand worm *Capitella sp*. They showed that this activity, although partially inhibited by fluoranthene, contributed to the burial of this PAH while increasing the fluoranthene content in the water column.

The previous analysis shows that studying a substance's effects cannot be approached without also studying the fate of this substance at the same time, as this is directly linked to its longevity in the different compartments of the ecosystem, and therefore the exposure conditions of the organisms (disappearance of the mother substance, degradation into more or less toxic metabolites, etc.). Barry et al. (1995) showed that endosulfan adsorbed on microalgae is less toxic for the daphnia ingesting these algae than dissolved endosulfan.

Whatever the case, the objective is to create conditions enabling the reproduction, even in a simplified version, of the portion of the ecosystem for which studying the fate and effects of a substance may allow a certain level of prediction, making it possible to extrapolate the results from the laboratory to a real site (Cairns and Cherry, 1993). To achieve this, a certain number of questions must therefore be answered concerning other scales, methods and the choice of variables in order to ensure the representativeness and the reproducibility of the system, and that pertinent indicators are used. Two approaches to laboratory microcosm tests have been developed. Laboratory microcosms can be entirely synthetic, i.e. elaborated from a reconstructed environment and sediment in which a definite number of species from the laboratory's cultures and breeding-stock are introduced. Synthetic or *gnotobiotic* (Taub, 1969) microcosms present the advantage of being simple and very reproducible, but they only mimic real ecosystems very schematically with a limited number of species (Barry and Logan, 1998). Laboratory microcosms based on natural communities from water and sediment taken *in situ* (Van Donk et al., 1995) are an alternative to gnotobiotic microcosms. These microcosms often provide results closer to what can be observed in natural ecosystems, but are harder to replicate (Barry and Logan, 1998).

In gnotobiotic microcosms, the ecosystem is entirely reconstructed from natural or synthetic sediment and water in which the organisms from the laboratory's culture and breeding stock are introduced. All the elements of the microcosm are therefore known if we disregard the microorganisms in the natural sediment, the diversity is limited, better controlled and reproducibility can be expected. This approach is not recent since Nixon used it as far back as 1969 with a synthetic microcosm of brackish waters, composed of a green alga (*Dunaliella viridis*), a micro-crustacean (*Artemia salina*), and non-identified bacteria from outside and then isolated (Nixon, 1969). Taub's SAMs (*Standardized Aquatic Microcosms*) are the best-known because they were subjected to a long adjustment period (Taub, 1969, 1989a, 1989b, 1997; Conquest and Taub, 1989; Taub et al., 1981, 1986, 1991) that led to a standardized protocol also used by other authors (Larsen et al., 1986; Stay et al., 1989; Rai and Mallick, 1993; Landis et al., 1993). A SAM is a 3-liter system comprising sand (+ chitin and cellulose), a synthetic nutritive medium, 10 species of algae, 5 animal species (daphnia, amphipods, ostracods, protozoans and rotifers), and bacteria (including *Enterobacter*). All the organisms come from the laboratory and are identified. Taub sought to demonstrate the effects of substances on important processes : photosynthesis, species dominance (algae and zooplankton), nutrient assimilation and recycling of organic matter, recovery, carry-over effects, and indirect effects. The protocol meets the following requirements : demonstration of ecological interactions inside (and between) the different trophic levels, capacity to highlight significant statistical differences in the populations and/or the abiotic factors caused by the toxic substance, sensitivity to numerous substances, repeatable and reproducible results, results that can be predicted and that are comparable to what happens *in situ* (Taub, 1989a). Although standardized, the implementation of this protocol is cumbersome, time- and energy-consuming because of the length of the test (63 days) and the need to maintain a large number of cultures and breeding stock in the laboratory. The sediment compartment is reduced to a medium for the development of a few organisms.

There are relatively few protocols for entirely gnotobiotic microcosms like SAMs. In many cases, the protocol is based on synthetic elements (for example, a culture medium for the water column) and organisms from breeding or else collected *in situ*. In certain cases natural sediment introduces microorganisms and even benthic macrofauna that complicate the systems and reduce their level of control. Mention can be made of the works by Pr. Gérard Blake's team (Savoie University). Merlin and Blake (1991) tested pentachlorophenol (PCP) in 2-liter microcosms simulating lentic ecosystems comprising a water column, lacustrine sediment and higher plants (*Pistia stratiotes* or water lettuce, floating species, rooted species like *Scirpus*, *Schoenoplectus lacustris*) and lower plants (filamentous algae like *Spirogyra* sp., *Gomphonema* sp.). Using only primary producers was justified by the herbicidal action of PCP. In the studies of Merlin *at al.* (1992) and Thiébaud et al. (1994) on the fate and the effects of dichloromethane, the microcosms contained artificial sediment (Fontainebleau sand), a synthetic water (modified Hoagland medium), immersed bryophytes (*Fontinalis antipyretica*), macrophytes (*Lemna minor, Groenlandia densa, Elodea canadensis*), mollusks (*Physa fontinalis*), micro-crustaceans (*Daphnia magna*) and a green alga (*Scenedesmus subspicatus*).

Other examples of microcosm test protocols that are not entirely gnotobiotic have been provided by the works of Sanchez and colleagues (Sanchez and Tarazona, 2002; Sanchez et al., 2004). Although the sediment was semi-artificial (sphagnum peat enriched sand), the water was natural (spring water). The microcosm was more a multi-species test since it only contained two (Sanchez et al., 2004) or three species (Sanchez and Tarazona, 2002). Besides, the daphnia were exposed free or caged, which enabled studying the toxic substance as a function of the type of exposure to the sediment.

In all these gnotobiotic or pseudo-gnotobiotic microcosm tests, the toxic substance is generally added rapidly (at most 1 week later, ASTM (1993)) and the colonization/ stabilization phase is shorter than in microcosms comprising natural communities. By contrast, the physico-chemical and biological variables monitored are not very different : pH, dissolved oxygen, electrical conductivity, NH_4^+, NO_3^-, NO_2^-, PO_4^{3-} contents, alkalinity, toxic content in the different compartments, chlorophyll pigments, density of the different organisms. Merlin and Blake (1991) also studied the primary photosynthetic and respiratory activities from foliar fragments and whole plants.

3. Conception and Application of Laboratory Microcosms in L.S.E./LEHNA-IPE and CEAEQ to Ecotoxicological Risk Assessment

3.1. General Description of the Different Types of Microcosms

Table 1 presents a summary of the protocols of the different types of microcosm tests that we have implemented. The smaller microcosm (volume: 2 L, figure 1), which is used most often, provides a schematic simulation of a portion of a lentic ecosystem, with the possibility of renewing the water column (static or semi-static test). The following parameters are monitored for 21 or 28 days: the physico-chemistry of the water column and sediment (including, when necessary, analytical monitoring of the substance(s) studied), the survival and reproduction of daphnia, algae and duckweed growth; the survival, growth and emergence of chironomids; and the survival and growth of amphipods.

The interest and advisability of developing a larger laboratory microcosm became apparent during specific studies (Clément et al., 2005a). A larger microcosm, for example, with a volume of 100 liters (figure 2), allows diversifying the ecosystem by adding new species (cladocera, gastropods, rooted plants), monitoring additional biological criteria (for example, amphipod reproduction, periphyton growth), and enhancing its ecological pertinence. The tests can be lengthened, from 30 days for a 2-liter microcosm to several months for a 100-liter microcosm. The size of the systems also enables the installation of traps and compartments that permit monitoring certain organisms exposed to specific conditions inside them (for example, exposed only to the water column, and/or inhibiting certain interactions with the environment or other organisms), while making the reconstructed ecosystem complex enough for the populations to evolve freely.

Table 1. Main parameters of 2-L, 100-L, and 180-L gravel pit microcosm assay protocols used by Clément and coll

	2 L microcosm	100 L microcosm	Gravel pit microcosm
Volume of water	2 L	100 L	180 L
Water depth	20 cm	40 cm	40 cm
Water type	synthetic medium	synthetic medium or enriched ground water	dechlorinated and enriched tap water
Renewal of water	no or intermittent	intermittent (10 L/week)	continuous (5 L/h)
Aeration	constant, air bubbling with Pasteur pipets		
Control sediment	formulated or natural	natural	natural
Mass of sediment	100 to 300 g	5 kg	72 kg
Illumination			
type	daylight fluorescent		
intensity	2000 lux at surface of water		
photoperiod	16 h light/day		
Equilibration time	7 days	ca 1 month	
Exposure time	21 or 28 days	90 days	
Organisms introduced			
primary producers	microalgae Ps duckweeds Lm	microalgae Ps duckweeds Lm, Sp rooted macrophytes Ec, Gd	microalgae Ps duckweeds Lm rooted macrophytes Ec
zooplankton	cladocerans Dm	cladocerans Dm, Cd, Sv	cladocerans Dm, Cd, Sv
benthos	chironomids Cr amphipods Ha	chironomids Cr amphipods Ha	chironomids Cr amphipods Ha
molluscs	no	snails Pa pond snails Ls	snails Pa pond snails Ls
Autochtonous organisms (origin : sediment)	/	ephemerid larvae oligochaetes periphyton	ephemerid larvae oligochaetes periphyton
Endpoints	survival, reprod Dm survival, growth Cr emergence Cr survival, growth Ha / /	 algal growth duckweed growth survival and growth of molluscs growth of rooted macrophytes	
Physico chemical parameters	water column : pH, O2, conductivity, TOC, anions and cations, contaminants (PAH, metals, ...) pore water : pH, TOC, anions and cations, contaminants		
Number of replicates	9 to 12	2	1
Sacrifices	on day 7, 14, 21 and 28	end of assay	

Ps : *Pseudokirchneriella subcapitata;* Lm : *Lemna minor;* Sp : *Spirodella polyrhiza;* Ec : *Elodea canadensis;* Gd : *Groelandia densa;* Dm : *Daphnia magna;* Cd : *Ceriodaphnia dubia;* Sv : *Simocephalus vetulus;* Cr : *Chironomus riparius;* Ha: *Hyalella azteca;* Pa: *Physa acuta;* Ls: *Limnaea stagnalis.*

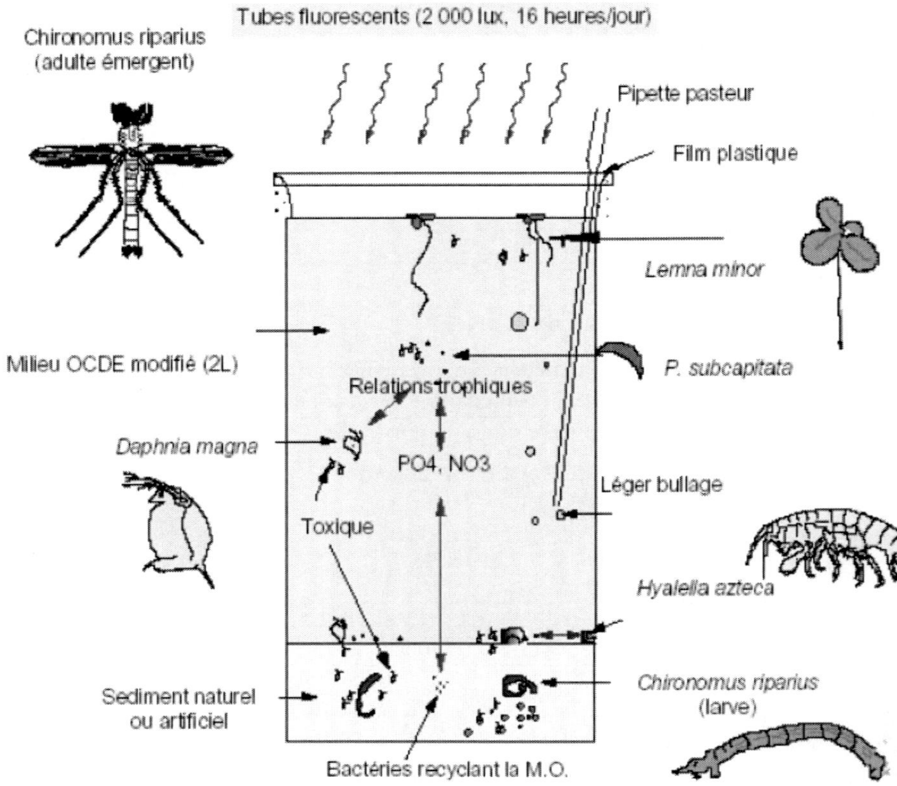

Figure 1. 2-L microcosm of Clément and coll.

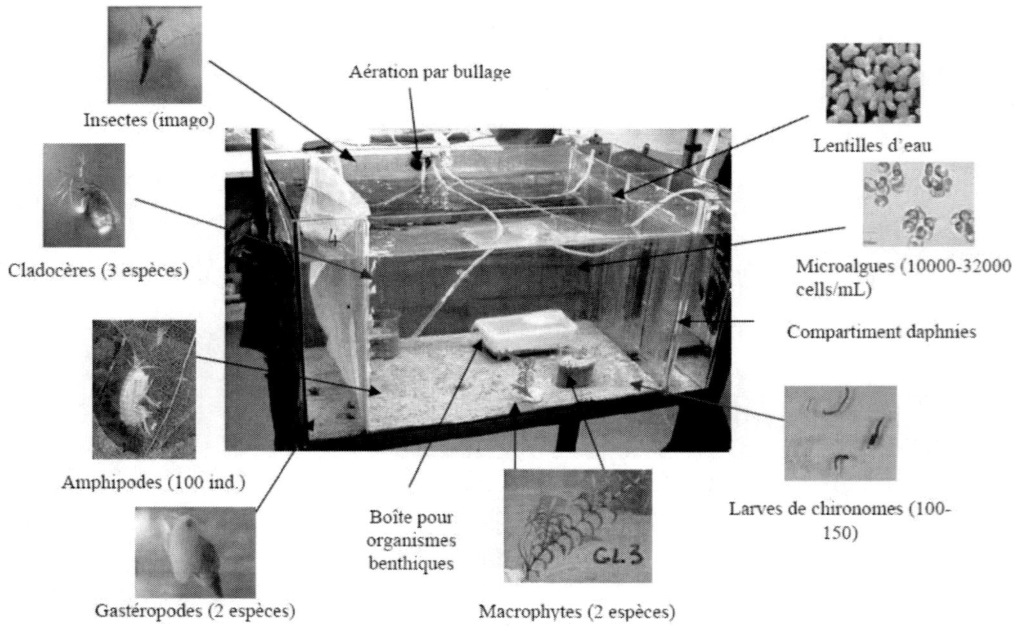

Figure 2. 100-L microcosm of Clément and coll. (Clément et al., 2005a).

Shortly afterwards, in the framework of the Novosol® research program driven by Solvay to evaluate the responses of reconstructed aquatic ecosystems to the immersion of sediment subject to inerting, we changed to a larger scale, that is to say to a more complex environment and a continuous water supply simulating the flow of groundwater. The aim was to simulate the immersion of sediment in a gravel pit before and after treatment. "Gravel pit" microcosms (figure 3) are composed of stainless steel tanks (2.30 m long by 50 cm wide by 1 m high) filled with random alluvial type fill in the center of which a free volume water of 270 L is held in a stainless steel compartment with a 1 mm mesh (dimensions: 1.5 m long by 30 cm wide by 60 cm high). The systems are continuously supplied (4 to 5 L/h) by the drinking water system. The water is dechlorinated and enriched in nitrogen and phosphorus before being percolated through the alluvial material, thus simulating the supply of a gravel pit by groundwater. The same clean sediment of lacustrine origin (Aiguebelette Lake, Savoie, France) as for the 2- and 100-L tests is introduced at the bottom of the central compartment. The organisms studied are the same as in 100-L microcosms and exposure time is 3 months.

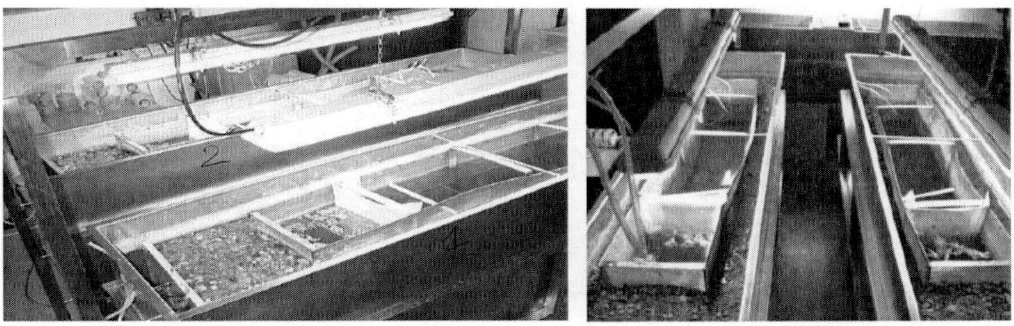

Figure 3. Gravel pit microcosm of Clément and coll. (Clément et al., 2010).

3.2. A Few Considerations on the Design of Microcosm Tests

The design of a microcosm test must satisfy numerous objectives and constraints. The microcosm must first offer, in the absence of toxic substances, favorable conditions for the development of the organisms introduced. This is based on the careful choice of the abiotic compartments, such as the sediment and overlying water, on certain functional conditions (temperature, oxygenation, water renewal, nutrition of the organisms), and finally on the species introduced, their density and predictable interactions with each other and with their environment. The fate and the effects of the pollutants introduced in the aquatic environment depend on the nature (texture, structure, composition) of the sediment and the physico-chemistry of the overlying water, partially influenced by the sediment as well as by the action of living organisms. The choice of these abiotic and biotic factors must therefore take into account the nature of the pollutants studied.

In laboratory microcosms, researchers use artificial aqueous media (Taub, 1989; Merlin et al., 1992; Rai and Mallick, 1993; Stay and Jarvinen, 1995; Sugiura, 1996) or filtered natural water enriched with nutriments (Brockway et al., 1979, 1984; Seguin, 2001). The use of these nutritive environments makes standardizing the tools easier. We used in some experiments natural water (either groundwater or from the drinking water system, after dechlorination and enrichment with nutriments) when water consumption was high (for the 100- and 200-L tests). The experiment showed that synthetic water is preferable because (i) it is possible to define a composition favorable for the correct development of all the organisms, and (ii) natural water is subject to variations of composition and contamination. At LEHNA-IPE we now mainly work with synthetic environments optimized for algal growth and the development of invertebrates (table 2). This environment is also propitious to the growth of duckweed. It is nevertheless possible to adapt its composition, notably its hardness and its EDTA content, which can significantly influence the bioavailability of trace metals.

Table 2. Composition of the synthetic medium used by Clément et coll. in 2-L microcosm assays

Solutions	Components	Stock solutions (g/l)	Concentration in medium (mg/l)
1	NH_4NO_3	1.1209	3.744 (1.31 mg N/l)
	KH_2PO_4	0.1268	0.424 (0.10 mg P/l)
2	$MgSO_4.7H_2O$	3.0	15.0
	$MgCl_2.6H_2O$	4.8	24.0
	$CaCl_2.2H_2O$	12.62	63.1
	$NaHCO_3$	25.05	125.25
4	$FeEDTANa2, 6H_2O$	0.1094	0.1094
	H_3BO_3	5.719	0.5719
	$MnCl_2.4H_2O$	0.721	0.0721
	$Na_2MoO_4.2H_2O$	0.126	0.0126
	LiCl	0.612	0.0612
	RbCl	0.142	0.0142
5	$SrCl_2.6H_2O$	0.304	0.0304
	NaBr	0.032	0.0032
	KI	0.0065	0.00065
	Na_2SeO_3	0.00438	0.000438
	NH_4VO_3	0.00115	0.000115
	$CoCl_2.6H_2O$	0.020	0.0020
	$CuSO_4.5H_2O$	0.033	0.0033
	$ZnSO_4.7H_2O$	0.026	0.0026
6	Vitamin B1 or Thiamin (HCl)	0.0750	0.0750
	Vitamin B12 or cyanocobalamin	0.0010	0.0010
	Vitamin H or Biotin	0.00075	0.00075

Unfortunately, there is no standard natural sediment available anywhere in large quantities. Composition diversity (total organic carbon, organic matter content, cationic exchange capacity, redox potential, pH, clay, silt and sand proportions) is extreme according to the data published by Suedel and Rogers (1994). Consequently, the choice of sediment is an unavoidable step in any biological test involving this matrix, be it for single-species or microcosm tests.

If the aim is to study the pollutants at a particular site, it seems logical to work with the sediment from that location, while bearing in mind that there is rarely a single sediment typical of a given site, and it might be necessary to first formulate a sampling strategy appropriate for the objective. Given the considerable heterogeneity of numerous sites, such an approach can fast become costly because of the number of samples necessary and the test to be performed.

Table 3. Composition of various formulated sediments used in laboratory single-species or microcosm assays

Sand	Kaolinite	Calcium carbonate	Alpha-cellulose	TetraMin	Peat moss	Spirulin	Meat extract	TOC	Reference
95.2%	0%	0%	4.80%	0%	/	/	/	2%	Clément & Cadier 1998
97.5%	0%	0%	2.44%	0-0.56%	/	/	/	1-1.3%	
65%[1]	30%	0.10%	4.75%	0.15%	/	/	/	2%	Triffault-Bouchet 2004
65%[1]	30%	0%	4.85%	0.15%	/	/	/	2%	Clément et al. 2004
65%[1]	30%	0.10%	4.75%	0.15%	/	/	/	2%	Verrhiest et al. 2001
77.2%[2]	19.9%	0.5 or 0.10%	2.40%	0.20%	0%	0%	0%	1%	Verrhiest et al. 2000
				0%	0%	0.075%	0.075%	1%	
			0%	0.20%	2.40%	0%	0%	1.3%	
62-66%	29-30%	0.10%	0.15%	0-4.75%	3.9-5.5-9.7%[5]	/	/	2-5.7%	Garric et al. 2001
75%[3]	20%	to reach pH7.0[4]		/	5%	/	/	2%	OECD 1998

[1] Loire sand sieved at 2 mm, particles between 500 μm and 2 mm, mean diameter 750 μm; [2] Fontainebleau sand, mean diameter of particles 150-210 μm; [3] industrial sand with more than 50% particles between 50 and 200 μm; [4] alternative to the use of peat moss; [5] brown or blond peat.

If the aim is to recreate inside the microcosm a sedimentary compartment possessing certain fundamental properties like the sequestration of pollutants or the capacity to develop benthic organisms, without reference to a specific geographic context, a reference natural sediment or an artificial sediment reconstructed from identified materials should be used. The formulation of artificial sediments is not standardized, but sand and clay often form the basis of their composition.

Table 3 displays the formulations used in the studies in which we have participated, as well as those accepted in the OECD guidelines for the chironomid test, similar to the first one. The compositions indicated are in % in relation to the total dry weight, the addition of a water column generally leads to a sediment with a water content of about 23% (compared to total wet weight).

Most authors performing tests on natural sediments choose to feed benthic organisms. In the absence of added food, these organisms must use the organic matter naturally present in the sediment, which can vary in quantity and quality and may not be sufficient to ensure a satisfactory level of survival or significant growth (Bonnet, 2000).

Table 4. Living species used in microcosms of Clément and coll

Type of species	Trophic level	Main habitat	Food sources	Microcosm
Green micro-algae	Primary producer	Water column	Autotrophic	all types
Floating plants	Primary producer	Water column/atmosphere	Autotrophic	all types 100 and 180-L
Rooted plants	Primary producer	Water column/sediment	Autotrophic	all types
Crustacean cladocerans	Primary consumers	Water column	Microalgae, POM, bacteria	all types 100 and 180-L
Crustacean amphipod	Primary consumers	Sediment surface	Microalgae, POM, bacteria	all types
Insect chironomid	Primary consumers	Sediment	Microalgae, POM, bacteria	all types
Mollusc gasteropod	Primary consumers	Ubiquitous	Periphyton, plants, duckweeds	100 and 180-L

In the case of artificial sediments, the incorporated organic matter is generally not very digestible (cellulose, humus, sphagnum peat). To provide organic matter adapted to the growth of organisms during tests on natural or artificial sediments, we use TétraMin®, which comes in the form of dried fish flakes. The same organic matter is used for breeding. TétraMin® is most often used as a daily dose during tests to avoid accumulation on the surface of the sediment and the degradation of the medium's quality due to the development of fermentation.

In line with the gnotobiotic systems we used as inspiration (Merlin and Blake, 1991; Merlin et al., 1992, 1993, 1994; Yosowidagdo, 1996; Conquest and Taub, 1989; Taub, 1989a, 1989b), we chose to study the response of a few species introduced simultaneously in all our 2- to 180-L microcosm tests. These species were most often bred, cultivated or acclimatized in the laboratory (table 4). The choice of the species fulfilled several requirements. It was first necessary to cover the main trophic levels: primary producers (algae and duckweed, rooted plants for larger microcosms), primary consumers (pelagic cladocerans and epibenthic (amphipods) and benthic (chironomids) invertebrates, as well as gastropod mollusks (pond snails, e.g. *Physa* sp., *Lymnaea* sp.) in 100- and 180-L microcosms), decomposers (microfauna and bacteria in the water column and above all the sediment that are either naturally present (natural sediment) or favored (artificial sediment)). Although the presence of these three trophic levels is not a sufficient condition to establish a true ecosystem, it is at least necessary to favor the interspecific relations and the functioning of material cycles to create an artificial though simplified ecosystem. Finally, it is therefore possible to obtain a large spectrum of sensitivities to toxic substances through the presence of plants, animals and prokaryotic organisms. The second aim is to work on organisms from lentic ecosystems specific to the compartments represented in the microcosm and therefore likely to highlight different paths of exposure: the organisms in the water column with the cladocerans, planktonic algae, duckweed, primarily exposed to the dissolved substances and particulates of the overlying water; the organisms of the sediment with amphipods and chironomids, as well as the bacterial compartment of the sediment. Finally the organisms used in the tests are generally easy to breed or cultivate and are classically chosen as test-organisms in laboratory bioassays, so that a great deal of toxicological data is available for each of them, facilitating the interpretation of the effects observed and the discussion of the results.

3.3. Examples of Microcosm Assay Applications for Ecotoxicological Risk Assessment

Bioavailability of PAHs

Microcosm bioassays were performed on sediments spiked with PAHs (mixture of fluoranthene, phenanthrene and benzo(k)fluoranthene; pyrene alone) in order to evaluate the risks for lentic ecosystems related to the presence of these ubiquitous organic contaminants in this sediment compartment (Verrhiest et al., 2000, 2001, 2002a, 2002b; Clément et al., 2005b). Main considerations on microcosm conception and a summary of main results are available in Clément (2012). These considerations relate to choice of sediment, sediment spiking, equilibration time following spiking, possible biodegradation during the assays, bioavailability of PAHs closely dependent on sediment characteristics (e.g. organic matter), type of light (photosensitivity and phototransformation of PAHs under UV light), sorption of PAHs to vessel walls and other materials, interactions in case of PAHs mixtures (synergy and additivity of effects), importance of multiple exposure (*via* water, particles, trophic chain). Results showed acute toxicity of PAHs in presence of light but also protection by organic matter for benthic organisms and probably biotransformation of PAHs by most organisms. The microbial compartment of sediment, analysed through enzymatic activities and cell counts, was not significantly impaired by PAHs, which is an important finding if one considers the importance of this compartment in ecosystem functioning. Finally, studying PAH toxicity should be carried out using spiked natural sediments rather than formulated sediments due to the importance of sediment composition and complexity in PAH fate and toxicity.

Eco-compatibility of MSW Incineration Bottom Ashes

In France, municipal solid waste incineration (MSWI) bottom ashes are widely used in road building. Leachates released from these materials may constitute a hazard for aquatic ecosystems. In order to assess this risk, a study of MSWI bottom ash leachates toxicity was carried out using single-species tests, 2-L laboratory microcosms (Triffault-Bouchet, 2004; Triffault-Bouchet et al., 2005a). A laboratory lysimeter was used to simulate the road embankment and produce MSWI bottom ash leachates. The single-species tests involved a green microalga (*Pseudokirchneriella subcapitata*), a floating plant (*Lemna minor*), and a cladoceran (*Daphnia magna*). The 2-L microcosms were those described in the present chapter. Results led to the following ranking of these bioassays based on organism sensitivity, in decreasing order: algae *Pseudokirchneriella subcapitata* > duckweed *Lemna minor* > daphnids *Daphnia magna*. In the microcosms, toxicity effects increased with leachate concentration, from 1.56 % to 8.0 %. Comparison between single-species tests and 2-L microcosm assays underlined the interest of the multispecies approach. Finally, MSWI bottom ashes were assessed using a simplified risk assessment procedure. Predicted environmental concentration was close to the concentration that caused first effects in microcosm. Recommendations were made for the reuse of MSWI bottom ashes as road embankment.

The study was continued with bigger microcosms (volume: 100 L) allowing a higher number of species exposed on a higher duration (60 days) (Clément et al., 2005a). The 100-L microcosms were glass tanks filled with a lake sediment and synthetic water, and inoculated with various organisms: microalgae, duckweeds, rooted macrophytes, cladocerans, pond snails, amphipods, and chironomids. In these assays organisms were exposed 50 to 60 days following leachate introduction. Using compartments and cages allowed to vary organisms exposure and limit interactions. The microcosms were contaminated several times so as to reach a final concentration of 6% (first assay) or 3% (second assay). The contamination led to increased salt (chlorides, sulfates, sodium and potassium) and metal (copper, chromium) concentrations. Among cladocerans, *Daphnia magna* was sensitive to contamination although recovery was observed, whereas *Simocephalus vetulus* and *Ceriodaphnia dubia* were not. Results regarding the amphipod *Hyalella azteca* were contradictory between both assays, perhaps due to change in major exposure. Chironomids emerged similarly in control and contaminated systems. Among pond snails, leachates impaired survival and grazing activity of *Lymnaea stagnalis*, whereas *Physa sp.* was not sensitive. The growth of all primary producers tested in the microcosms (suspended microalgae and periphyton, floating and rooted plants) was not impaired by MSWI bottom ash leachates. Copper might have explained most of observed effects. Due to the fact that concentrations studied here seem to be an overestimation of real field concentrations, and that MSWI bottom ash leachates characteristics decrease with time, it was concluded that the risks for lentic aquatic ecosystems would be minimal.

Immersion of Dredged Sediments

One possible fate of dredged sediments is disposal in a submerged gravel pit, but this practice may be a source of contamination of the aquatic environment (Clément et al., 2004). In order to assess the risks for the aquatic ecosystem, 2-L microcosm assays were conducted in the laboratory on three dredged sediments from a french canal. Pelagic organisms (microalgae, daphnids, duckweeds, pond snails) and benthic organisms (amphipods, chironomids) were introduced into the microcosms. Survival and growth of organisms were monitored during 28 days. Genetic damage was measured weekly in pond snail hemocytes through the comet assay. The sediments contained high levels of metals (up to 740 mg copper/kg and 1220 mg zinc/kg), but were moderately contaminated by PAHs (total PAH content < 10 mg/kg) and PCBs (total PCB content < 0.6 mg/kg). Only moderate effects on the biota were observed, except for genetic damage on pond snail hemocytes which was significant in all microcosms. The absence of pronounced toxic effects on organisms was not expected since concentrations of some sediment toxicants were above threshold values reported in the literature, especially for metals in two sediments, and for PAHs in the three sediments. Speciation, influence of organic matter and presence of sulfides might explain the low toxicity observed. By contrast, assessment of DNA damage revealed to be more sensitive and enabled a ranking of sediments coherent with their toxic load. In the present work, the microcosm protocol previously set up with a formulated sediment appears to be fitted to natural sediments.

In another research focussed on filling submerged gravel pits with treated sediments (Novosol® treatment consisting in neutralizing metals by complexation with orthophosphoric acid and removing organic pollutants by calcination) in order to minimize hazards for the environment, we assessed the efficiency of this physico-chemical treatment in a scenario of sediment immersion into experimental laboratory gravel pits (LGP) (Clément et al., 2010a). A 180-L water compartment with continuous renewal of water was set up in each system so as to simulate the gravel pit, and various living organisms were introduced. Following a period of colonization and stabilization, raw and treated sediments were introduced into two different LGPs, and the fate and effects of pollutants were studied during the period of deposition and post-deposition. The sediment treatment had positive effects on survival and development of benthic populations and reproduction of pond snails but the introduction of the treated sediment was followed by an increase in salinity (phosphates, sulphates) and a peak of hexavalent chromium at concentrations above drinkability limits and likely to have impaired invertebrate populations of the water column. The results of this study suggested that discharge of contaminated sediments at a high solid:liquid ratio (1:10) in gravel pits or equivalent aquatic ecosystems may have only limited effects on biota and ground water quality. It was nevertheless underlined that the Novosol® treatment should be improved so as to increase efficiency of oxidised chromium complexation during the phosphatation step.

Ecotoxicological Assessment of Road Runoff Residues

Road runoff residues are sediments accumulated in basins and ponds built in the vicinity of roads and motorways to collect runoff waters. These materials are valorised in road and civil works building, or stored into dumps when they are highly contaminated. When valorised, they may be drained by rain water and the corresponding leachates may reach close aquatic ecosystems. We assessed the ecotoxicological risks for lentic aquatic ecosystems receiving the leachates of road runoff residues, either raw or treated, through granulometric sieving (Clément et al., 2010). Ecotoxicological single-species tests and microcosm assays were carried out on 15 leachates from sediments collected in various locations (13 raw materials and 2 valorizable granulometric fractions). All leachates showed at least one metal concentration above predicted no effect concentration (PNEC). No acute toxicity was found at concentration <20% leachate (v:v). Among the 15 non filtered leachates, only two were toxic to some organisms of microcosms, the others were not toxic at the tested concentration of 10%. In this study, the use of microcosm assays was particularly relevant, as these microcosms allow to better represent the complex environment (water column and sediment) and to measure effects on rather long exposure durations.

Ecotoxicological Impacts of a Lake Restoration in Québec

Eutrophication and contaminated sediments (*e.g.* nutrients, metals, etc.) are preoccupying issues in aquatic ecosystems and water resource protection. Saint-Augustin Lake (Quebec, Canada) is affected by both problems in addition to the presence of de-icing salts. This small shallow lake has low renewal time and receives an intensive phosphorus transfer from bottom sediments to the water column. Since 2001, the lake presents cyanobacteria blooms every summer ($P > 70$ µg/L). A restoration technique for this eutrophic lake was assessed for its ecotoxicological impacts. The technique combined the application of alum (aluminum sulphate, $Al_2(SO_4)_3$) and an active capping (calcite and sand). A battery of monospecific assays (growth of *Pseudokirchneriella subcapitata* and *Lemna minor*, survival and reproduction of *Daphnia magna*, survival, growth and reproduction of *Hyallela azteca* and survival, growth and emergence of *Chironomus riparius*) and a multi-species laboratory assay

(2-L microcosm) were used to assess the impacts of the application of alum alone and of the complete technique. Overall, both methodological approaches gave similar results. Microcosm assays highlighted more subtle effects, specially on daphnid and chironomids. Indeed, without treatment, the water and sediment of the lake (i) stimulated the growth of the primary producers (algae, duckweed, cyanobacteria) in relation with TP concentrations, and (ii) impaired the development of the primary consumers (daphnids, amphipods, chironomids). The proposed technique was effective in reducing TP concentrations by 60 %, and consequently, the growth of primary producers. The daphnid reproduction and the chironomid emergence were slightly improved in the treated microcosm systems. Parameters measured on amphipods were not modified, whatever the treatment. Finally, the new environmental conditions were more beneficial for the primary consumers. Recommendations were given for *in situ* studies as well as for a possible application of the restoration technique in the Saint-Augustin Lake.

3.4. Optimisation of the 2-liter Laboratory Microcosm Protocol

The 2-L microcosm test protocol (Clément and Cadier, 1998) enables evaluating the response of different organisms exposed simultaneously, under conditions where they interact with each other and with the environment, to one or more toxic substances introduced *via* the water column or the sediment. These 2-L microcosm tests seem well-suited for ecological risk assessment linked to the deposit of contaminated (Babut et al., 2002; Clément et al., 2004; Triffault-Bouchet, 2004; Triffault-Bouchet et al., 2005b) and treated sediments in lentic systems (Cauzzi, 2007), once the deposit has stabilized (non-disturbed sediment under water). Although the results on benthic organisms are not substantially different from those obtained with single-species tests, microcosm tests allow demonstrating that pelagic organisms (daphnia, *Lymnaea* sp.) exposed in the water column overlying sediments present high sensitivity in certain cases. This appears due more to exposure to pollutant particles in the sediment than to the contents of pollutants dissolved in the water column (Triffault-Bouchet, 2004).

When comparisons with results obtained with single-species tests are possible, they often show different responses (algae are more sensitive to the leachates of bottom ash from municipal solid waste incineration in single-species tests, whereas the amphipod *H. azteca* is less sensitive to the leachates of this ash when it can take refuge in the sediment), probably linked to the very different exposure conditions and the transformation of toxic substances by the system (partition into different compartments, hydrolysis, evaporation in certain cases, photodegradation, biodegradation, adsorption onto particles, etc.).

It is important to ensure that exposure time is long enough, which is possible with microcosm tests, to take into account both chronic effects and the attenuation of effects, the degradation of the substances tested and the modification of their bioavailability (the attenuation of the effect of leachates of bottom ash from municipal solid waste incineration on *D. magna* due to the transformation of copper in the medium (reduction of its bioavailability)), and the degradation of PAHs tested by Verrhiest et al. (2002).

Microcosm tests enable studying the fate and bioavailability of pollutants by the presence of the main compartments influencing them, namely the water column, sediment and biota. That is what we have done with hydrophobic compounds like PAHs. The presence of several exposure compartments has led us to highlight the different responses of daphnia exposed to a mixture of PAHs in microcosms compared to the same daphnia exposed solely to water. The contamination level of the overlying water does not always explain the effects on the organisms in the water column also exposed to the sediment. We were also able to verify this with multi-contaminated sediments. Additionally, these different compartments interact with each other and thus influence the fate of the toxic substances: sediment contributes dissolved and colloidal organic matter and particles to the overlying water column thereby influencing the fate of hydrophobic substances in it, the bioturbation and burial activity of benthic organisms modifies the partition of the toxic substances in the sediment, etc.

Microcosms allow studying a large array of effects at different trophic levels, from bacteria (studies on PAHs) to macro-invertebrates, including primary producers, and highlighting variable sensitivities to toxic substances and materials tested with different organisms, as well as within the same group (*Daphnia magna, Ceriodaphnia dubia* and *Simocephalus vetulus* exposed to leachates of bottom ash from municipal solid waste incineration).

We have been able to highlight certain interactions inside microcosms: for example lower toxicity of pyrene to daphnia in the presence of algae, interactions between algae and daphnia as a function of the content of leachates of bottom ash from municipal solid waste incineration, the bioturbation of oligochaetes resulting in the burial of contaminated sediment. Despite the qualities and advantages presented above, a recurrent problem of microcosm tests is the variability of the data obtained, which harms the detection capacity and the sensitivity of the effect criteria taken into account (Caquet et al., 2001). This variability is inherent to microcosm tests. Due to the self-organization that takes place, each microcosm develops unique properties (Beyers and Odum, 1993), even in the case of relatively simple microcosms like 2-L microcosms. As we have observed, introducing one or more toxic substances leads to additional variability since it increases the number of degrees of freedom of the system through an additional variation factor. The analysis of such tests should be done while bearing in mind that the variability present is nothing more than the expression of the possible evolutions of the micro-ecosystems that can be initiated by the effect of toxic stress, in the same way that a population of identical individual daphnia can present varying responses when exposed to a toxic substance. Although the variability within microcosms is generally greater for biological parameters than for physico-chemical ones, as demonstrated by Triffault-Bouchet (2004) for 2-L microcosms and Caquet et al. (2001) for mesocosms at the INRA biological station of Rennes, certain criteria lead to better results. Based on the analysis of all the biological data for 2-L microcosm tests performed in this report, and confining ourselves to the variability existing within the tests, we opt for the survival and the dry weight of chironomid larvae, the day of first emergence and the time it takes to obtain 50% of emergences, and the survival and dry weight of the amphipods at 21 and 28 days. For algae and daphnia, variability increases significantly and leads to a lack of sensitivity of the parameters affected (algal growth, and the survival and reproduction of daphnia).

The question of variability has been dealt with in depth recently (Delhaye, 2012) in view to optimizing the protocol of 2-L microcosm tests. Delhaye studied sources of variability in microcosm assays and proposed improvements to the protocol in order to reduce this variability. Modifications were first brought to the basic static protocol so as to improve organisms fitness: increase of water nutrient contents to support a higher growth of algae and duckweeds and hence enhance daphnid development, increase of water hardness to a value fitted to daphnids and amphipods, addition of vitamins. The highest improvement of microcosm conditions and reduction of variability was obtained with a continuous renewal of water. Variability was reduced for conductivity and hardness, and renewal led to stabilisation of most physico-chemical parameters. Although renewal of water led to loss of microalgae in the outlet, these ones were able to grow in the system and feed the daphnids, and variability was lower than in the static assay (CV = 19%). Improvements were also noticed for duckweeds, daphnids, amphipods and chironomids, due to increase of fitness and decrease of variability. In experiments with cadmium as toxicant, renewal and continuous input of cadmium in the inlet water allowed a stabilisation of water cadmium content. Due to the reduction of variability, it was possible to detect statistical significant effects at concentrations lower than in the static test, especially for the duckweeds. It should nevertheless be kept in mind that exposure to cadmium was quite different in both systems, with decreasing concentrations of Cd in water in the static test versus constant concentrations in the dynamic test.

3.5. Perspectives: Modelling as an Aid to Improve Interpretation of Microcosm Bioassays Results

The aim of ecotoxicological assays is to relate exposure concentrations and effects of a given chemical (or a mixture of chemicals, or complex environmental samples such as effluents or leachates which may contain confounding factors such organic matter and nutrients). In a complex system such as the present microcosm (two phases: water and sediment, and five species), organisms may be exposed in multiple ways (water, sediment, food, suspended particles,...); hence it is difficult to relate observed effects and a toxicant concentration external to the organisms, all the more this concentration can vary in time and space for various reasons (degradation, sorption onto particles, etc.).

Moreover, conversely to single-species tests where the variations of the response can be simply linked to the presence of the studied toxicant and compared to controls, the differences observed in microcosms may result from indirect as well as direct effets, or a combination of both. For example, when an herbicide reduces algal growth, it may also impair growth and reproduction of daphnids which ingest the algae, while this herbicide has no direct effects on the invertebrates. The sole comparison of contaminated and control systems shows indeed a global effect but fails to produce more precise information. In order to improve interpretation of effects observed in microcosms, the modelling of the functioning of microcosm could allow a more detailed explanation of observations and the testing of various assumptions on the mechanisms which control organism responses and their interactions in the presence of toxicants (*in silico* analyses). Due to the complexity of the microcom, it may be necessary to use dynamic modelling tools based on a system of coupled differential equations. As a matter of fact, it is possible to describe the main components and their interactions by mathematical equations, as soon as the main processes involved have been identified. The mathematical model thus allows to describe the evolution of the main key variables of the system. Moreover, if it is possible to couple this functioning model to individual centered effect models such as the mechanistic model DEBtox (Kooijman and Bedaux, 1996) or any other model allowing the introduction of stress functions on biological variables monitored (growth, reproduction, survival, ...), it then becomes possible to obtain a complete model able to represent the evolution of the system in the presence of a toxicant. A future work is going to start through a PhD thesis on the modelling of the responses of the various organisms of the microcosm exposed to a toxicant such as the cadmium, upon which numerous literature data are available and which can easily be analysed in our lab (LEHNA-IPE). The research will start with the description of the conceptual model of the microcosm taking into account all organisms and with the study of the way of coupling a dynamic model to an effect model. A first study will be performed on the compartment microalgae/daphnids, taking into account the main processes involved in the evolution of both populations, on the basis of recent experimental work already carried out, and completing this work if necessary. This approach will be extended to the other organisms of the microcosm.

Preliminary studies have been carried out to build the foundation of this new research topic. Up to now, a first study of an ENTPE student (Velluet, 2009) was carried out with the dynamic modelling software VENSIM, focussing on the response of microalgae population to various factors (growth, sedimentation, grazing by daphnids) under different conditions of water renewal (no renewal or continuous renewal), which allowed to partially define the conceptual model, the equations of the processes, and to evaluate the possibilities of simulation. Concomitantly, a collaboration between LBBE (Laboratoire de Biométrie et Biologie Evolutive, UMR 5558, CNRS, Université Lyon 1) and LEHNA led to modellise the effects of cadmium on daphnids exposed in microcosms in two conditions of renewal (Billoir et al., 2011, 2012). Three Master internships (Gauci, 2011; Orias, 2012; Herbach, 2012) have resulted in improving the model of the microalgae/daphnids compartment.

CONCLUSION

We go along with Cairns *at al.* (1992) in asserting the need to develop test protocols capable of more precise and pertinent assessment of the concentrations of toxic substances presenting a risk for ecosystems, and which can produce information to predict possible consequences *in situ*. The response of natural ecosystems cannot be predicted on the basis of the responses of a few species tested separately in a laboratory, because the behavior of the former depends on the processes at work at different levels of biological organization: those of the individuals, the populations, the communities and the ecosystem. Standard tests are used to measure the responses of a few species to a toxic substance, but not the effects on the dynamics of communities and ecosystems. Since it is probable that no species is systematically more sensitive than another, it is preferable to use several species from different taxonomic and functional levels. However, even with batteries of tests, this does not enable evaluating the effects on higher processes (biomagnification, competition, etc.). The biggest problem with single-species tests is their poor environmental realism: absence of competitors, predators, parasites that can weaken individuals, organisms free of infection and with unlimited nutrition, and conditions that, in a natural environment, can increase sensitivity to stress. Given the inherent uncertainty of these parameters, the application of a security factor can lead to overestimating or underestimating the risk, and therefore to an additional environmental cost. The use of microcosm tests appears to improve things, provided that the uncertainty on the results is not too great and that they can be extrapolated to real ecosystems. The efforts made to enhance the realism of experimental conditions (type of environment, presence of sediment, etc.) and to expose different species simultaneously over sufficiently long periods should, *a priori,* improve prediction of the effects that would be observed *in situ*. The optimization of the 2-L microcosm test protocol by enhancing the development conditions of the organisms introduced and the flow-through renewal of the water column increases the statistical power of the test while lessening the variability of the responses of the test-organisms and increasing their fitness. Works in progress on 2-L microcosms are aimed at developing a corpus of knowledge in ecotoxicological modelling that should, on the one hand, improve the interpretation of the effects observed on the organisms of the microcosm and, on the other hand, provide effect thresholds for these organisms under dynamic conditions (i.e. biotic and abiotic compartments evolving under the combined effect of the toxic substance and the development of the organisms).

REFERENCES

ASTM (1993). Standard practice for Standardized Aquatic Microcosms: Fresh Water, ASTM Standards on Aquatic Toxicology and Hazard Evaluation, E1366-91, pp 422-456.

Babut M., Perrodin Y., Bray M., Clément B., Delolme C., Devaux A., Durrieu C., Garric J., Vollat B., Bécart, D., Charrier C. (2002). Évaluation des risques écologiques causés par des matériaux de dragage: proposition d'une approche adaptée aux dépôts en gravière en eau. *Rev. Sci. Eau* 15: 615-639.

Barry M.J., Logan D.C. (1998). The use of temporary pond microcosms for aquatic toxicity testing: direct and indirect effects of endosulfan on community structure, *Aquat. Toxicol.* 41: 101-124.

Barry M.J., Logan D.C., Ahokas J.T., Holdway D.A. (1995). The effects of algal food concentration on toxicity of two agricultural pesticides to *Daphnia carinata*, *Ecotoxicol. Eviron. Safety* 32: 273-279.

Beyers, Robert J. and Howard T. Odum, 1993. Ecological Microcosms (Springer Advanced Texts in Life Sciences). Springer Verlag. ISBN: 0387979808

Billoir E., Delhaye H., Clément B., Delignette-Muller M.L., Charles S. (2011). Bayesian modelling of daphnid responses to time-varying cadmium exposure in laboratory aquatic microcosms, *Ecotoxicology and Environmental Safety* 74: 693–702.

Billoir E., Delhaye H., Forfait C., Clément B., Triffault-Bouchet G., Charles S., Delignette-Muller M.L. (2012). Comparison of bio-assays with different exposure time patterns: the added value of dynamic modelling in predictive ecotoxicology, *Ecotoxicology and Environmental Safety* 75: 80–86. Bonnet C. (2000). Développement de bioessais sur sédiments et applications à l'étude, en laboratoire, de la toxicité de sédiments dulçaquicoles contaminés, spécialité «Toxicologie de l'Environnement» Université de Metz, 326 pages.

Brockway D.L., Hill J., Maudsley J.R., Lassiter R.R. (1979). Development, replicability and modeling of naturally derived microcosms, *Intern. J. Environ. Studies* 13: 149-158.

Brockway D.L., Smith P.D. and Stancil F.E. (1984). Fate and Effects of Atrazine in Small Aquatic Microcosms, *Bull. Environ. Contam. Toxicol.* 32: 345-353.

Cairns J.Jr., Cherry D.S. (1993). Freshwater Multi-Species Test Systems, 101-116, Handbook of Ecotoxicology, Volume one, P. Calow (Ed.), Blackwell Scientific Publications, 478 pages.

Cairns J.Jr., McCormick P.V., Belanger S.E. (1992). Ecotoxicological testing: small is reliable, *J. Environ. Pathol. Toxicol. Oncol.* 11: 247-263.

Calow P. (1993). General principles and overview *in* Handbook of ecotoxicology, Calow P. (ed.), Blackwell Scientific Publications, Oxford, 1-5.

Caquet T., Lagadic L., Jonot O., Baturo W., Kilanda M., Simon P., Le Bras S., Echaubard M., Ramade F. (1996). Outdoor experimental ponds (mesocosms) designed for long-term ecotoxicological studies in aquatic environment, *Ecotoxicol. Environ. Safety* 34: 125-133.

Caquet T., Lagadic L., Monod G., Lacaze J.C., Couté A. (2001). Variability of physico-chemical and biological parameters between replicated outdoor freshwater lentic mesocosms, *Ecotoxicology* 10: 51-66.

Cauzzi N. (2007). Evaluation de l'éco-compatibilité de sédiments contaminés, traités et non traités par un procédé physico-chimique, dans le cadre d'un scénario de dépôt en gravière. Etude en microcosmes aquatiques. Thèse de doctorat de l'INSA de Lyon, Ecole doctorale Chimie de Lyon, Laboratoire des Sciences de l'Environnement de l'ENTPE, 11 juillet 2007, 341p.

Clément B. (2006). Apports des essais en microcosmes aquatiques lentiques de laboratoire à l'évaluation écotoxicologique des polluants. Rapport d'HDR INSA de Lyon-Université Lyon 1, 279 p..

Clément B. (2012). Bioavailability of Polycyclic Aromatic Hydrocarbons Studied Through Single-Species Ecotoxicity Tests and Laboratory Microcosm Assays, Organic Pollutants Ten Years After the Stockholm Convention - Environmental and Analytical Update, Tomasz Puzyn and Aleksandra Mostrag-Szlichtyng (Ed.), ISBN: 978-953-307-917-2, InTech, DOI: 10.5772/33304. Available from: http://www.intechopen.com/books/organic-pollutants-ten-years-after-the-stockholm-convention-environmental-and-analytical-update/bioavailability-of-polycyclic-aromatic-hydrocarbons-studied-through-single-species-ecotoxicity-tests.

Clément B., Cadier C. (1998). Development of a new laboratory freshwater/sediment microcosm test, *Ecotoxicology* 9: 279-290.

Clément B., Cauzzi N., Godde M., Crozet K., Chevron N. (2005b). Pyrene toxicity to aquatic pelagic and benthic organisms in single-species and microcosm tests, *J. Polycyclic Aromatic Compounds* 25: 271-298.

Clément B., Devaux A., Perrodin Y., Danjean M., Ghidini –Fatus M. (2004). Assessment of sediment ecotoxicity and genotoxicity in freshwater laboratory microcosms *Ecotoxicology* 13: 323-333.

Clément B., Raevel V., Renard O. (2010b). Ecotoxicological assessment of road runoff residues for aquatic surface ecosystems in a scenario of re-use, *Journal of Soils and Sediments*, 10: 1255–1266.

Clément B., Triffault-Bouchet G., Lottmann A., Carbonel J. (2005a). Are percolates released from solid wastes incineration bottom ashes safe for lentic ecosystems ? A laboratory ecotoxicological approach based on 100-litre indoor microcosms, *Aquatic Ecosystem Health Management* 8: 427-439.

Clément B., Vaille G., Moretto R., Vernus E., Abdelghafour M. (2010a). Effects of a physico-chemical treatment of a dredged sediment on its ecotoxicity after discharge in laboratory gravel pit microcosms, *Journal of Hazardous Materials* 175: 205–215.

Clément, B., Cauzzi, N., Godde, M., Crozet, K. & Chevron, N. (2005b). Pyrene toxicity to aquatic pelagic and benthic organisms in single-species and microcosm tests, *Polycyclic Aromatic Compounds* 25: 271-298.

Conquest L.L., Taub F.B. (1989). Repea tability and reproducibility of the Standardized Aquatic Microcosm: statistical properties, *Aquatic Toxicology and Hazard Assessment: 12th Volume*, ASTM STP 1027, American Society for Testing and Materials, Philadelphia, 159-177.

Delhaye H. (2012). Développement d'un bioessai de toxicité chronique en microcosme aquatique de laboratoire et évaluation de l'outil au travers de l'étude des effets du cadmium, rapport de thèse de l'INSA de Lyon, Ecole doctorale Chimie de Lyon, 319 pages.

Fleeger J.W., Carman K.R., Nisbet R.M. (2003). Indirect effects of contaminants in aquatic ecosystems, *Sci. Tot. Environ.* 317: 207-233.

Garric J., Bray M., Vollat B., Clément B., Devaux A., Delolme C., Durrieu C., Bedell J.P. (2001). Evaluation écotoxicologique de sédiments contaminés ou de matériaux de dragage. (II). Présentation des méthodes d'essai, rapport Cetmef/Drast et VNF, avril 2001, 46 pages.

Gauci F-X (2011). Contribution à la modélisation du fonctionnement d'un microcosme de laboratoire - Elaboration d'un modèle algues-daphnies en milieu microcosme, rapport de Master Sciences de l'Environnement Industriel et Urbain, 53 pages.

Giddings J.M., Biever R.C., Helm R.L., Howick G.L., Denoyelles Jr. F.J. (1994). The fate and effects of guthion (azynphos methyl) in mesocosms, In Aquatic Mesocosm Studies in Ecological Risk Assessment, R.L. Graney, J.H. Kennedy, J.H. Rodgers, Eds, SETAC Special Publications Series, Lewis Publishers.

Heimbach F., Pflueger W., Rätte H.T. (1992). Use of small artificial ponds for assessment of hazards to aquatic ecosystems, *Environ. Toxicol. Chem.* 11: 27-34.

Herbach U. (2012). Interaction algues-daphnies au sein d'un microcosme: modélisation dynamique, analyse de sensibilité et optimisation non linéaire, rapport de Magistère de Mathématiques de Rennes, 1ère année, 60 pages.

Kimball K.D., Levin S.A. (1985). Limitations of laboratory bioassays: the need for ecosystem-level testing, *Bioscience* 35: 165-171.

Kooijman, S., Bedaux, J. (1996). The Analysis of Aquatic Toxicity Data. VU University Press, Amsterdam.

Landis W.G., Matthews R.A., Markiewicz A.J., Matthews G.B. (1993). Multivariate analysis of the impacts of the turbine fuel JP-4 in a microcosm toxicity test with implications for the evaluation of ecosystem dynamics and risk assessment, *Ecotoxicology.* 2: 271-300.

Larsen D.P., Denoyelles F. Jr., Stay F., Shiroyama T. (1986). Comparisons of single-species, microcosm, and experimental point responses to atrazine exposure, *Environ. Toxicol. Chem.* 5: 179-190.

Lay J.P., Peither A., Jüttner I., Weiss K. (1993). *In situ* pond mesocosms for ecotoxicological long-term studies, *Chemosphere* 26: 1137-1150

Madsen S.D., Forbes T.L., Forbes V.E. (1997). Particle mixing by the polychaete *Capitella* species 1: coupling fate and effect of a particle-bound organic contaminant (fluoranthene) in a marine sediment, *Marine Ecology Progress Series* 147: 129-142.

Merlin G., Blake G. (1991). Absorption des pesticides par les macrophytes aquatiques: effets comparés sur la photosynthèse, cas du PCP, *Verh. Internat. Verein. Limnol.*, 24: 2282-2288.

Merlin G., Kilanda M.M., Blake G. (1993). Evaluation of ecotoxicological impact of detergent effluents using laboratory microcosms, *Fresenius Envir. Bull.* 2: 185-192.

Merlin G., Thiébaud H., Blake G., Sembiring S., Alary J. (1992). Mesocosms and microcosms utilization for ecotoxicity evaluation of dichloromethane, a chlorinated solvent, *Chemosphere* 24: 37-50.

Merlin G., Yosowigdagdo R., Thiébaud H., Blake G., Benoit-Guyod J.-L. (1994). Application of aquatic laboratory microcosms as experimental exposure assessment models: comparison between methylene chloride and xylylxylene. SETAC Europe 4th Annual Meeting, Brussels, April 11-14th 1994.

Nixon S.W. (1969). A synthetic microcosm, *Limnol. Oceanogr.* 14: 142-145.

OCDE (1998). OECD Guidelines for the Testing of Chemicals - Proposal for a new guideline – Chironomid Toxicity Test Using Spiked Sediment, Draft Document, May 98, 14 pages.

Orias F. (2012). Contribution expérimentale à la modélisation d'un microcosme algues-daphnies en conditions normale et contaminée (cadmium), rapport de Master Ecologie Biodiversité Evolution, Université Paris Sud XI, 47 pages.

Pratt J.R., Niederlehner B.R., Bowers N., Cairns J.Jr. (1987). Prediction of permissible concentrations of copper from microcosm toxicity tests, *Toxicity Assessment: An International Quaterly* 2: 417-436.

Rai L.C., Mallick N. (1993). Heavy metal toxicity to algae under synthetic microcosm, *Ecotoxicology* 2: 231-242.

Sánchez P., Alonso C., Fernández C., Vega M.M., García M.P., Tarazona J.V. (2004). Evaluation of a Multi-species Test System for Assessing Acute and Chronic Toxicity of Sediments and Water to Aquatic Invertebrates - Effects of Pentachlorophenol on *Daphnia magna* and *Chironomus prasinus*, *J Soils & Sediments* (OnlineFirst): 1 - 6.

Sánchez P., Tarazona J.V. (2002). Development of a multispecies system for testing reproductive effects on aquatic invertebrates. Experience with *Daphnia magna, Chironomus prasinus* and *Lymnaea peregra, Aquat. Toxicol.* 60: 249-256.

Seguin F. (2001). Effets comparés de deux herbicides, l'atrazine et le nicosulfuron, sur un écosystème aquatique expérimental. Chapitre 1: Mésocosmes aquatiques, éléments de caractérisation, pp 1-22. Thèse Université Toulouse III.

Stay F.S., Flum T.E., Shannon L.J., Yount D.J. (1989). An assessment of the precision and accuracy of SAM and MFC microcosms exposed to toxicants, *Aquatic Toxicology and Hazard Assessment:* 12th volume, ASTM STP 1027, U.M. Cowgill and L.R. Williams, Eds., pp 189-203.

Stay F.S., Jarvinen A.W. (1995). Use of microcosm and fish toxicity data to select mesocosm treatment concentrations, *Arch. Environ. Contam. Toxicol.* 28: 451-458.

Suedel B.C., Rodgers Jr. J.H. (1994). Development of formulated reference sediments for freshwater and estuarine sediment testing, *Environ. Toxicol. Chem.* 13/7: 1163-1175.

Sugiura K. (1996). The use of an aquatic microcosm for pollution effects assessment. *Wat. Res.* 30: 1801-1812. Brockway D.L., Hill J., Maudsley J.R., Lassiter R.R. (1979). Development, replicability and modeling of naturally derived microcosms, *Intern. J. Environ. Studies* 13: 149-158.

Taub F. (1969). A biological model of a freshwater community: a gnotobiotic ecosystem, *Limnol. Oceanogr.* 14: 136-141.

Taub F.B. (1989a). Standardized Aquatic Microcosm. Development and Testing. *Aquatic Ecotoxicology: Fundamentals, Concepts and Methodologies*, vol.2, Boudou A., Rybeyre F. (Eds) CRC Press, Boca Raton, Florida, pp 47-91.

Taub F.B. (1989b). Standardized Aquatic Microcosms, *Environ. Sci. Technol.* 23: 1064-1066.

Taub F.B. (1997). Unique information contributed by multispecies systems: examples from the Standardized Aquatic Microcosm, *Ecological Applications* 7: 1103-1110.

Taub F.B., Harrass M.C., Hartmann H.J., Kindig A.C., Read P.L. (1981). Effects of initial algal density on community development in aquatic microcosms, *Verh. Internat. Verein. Limnol.* 21: 197-204.

Taub F.B., Kindig A.C., Conquest L.L. (1986). Preliminary Results of Interlaboratory Testing of a Standardized Aquatic. *Community Toxicity Testing*, ASTM STP 920, Cairns, Jr., pp 93-120.

Taub F.B., Kindig A.C., Meador J.P., Swartzman G.L. (1991). Effects of "seasonal succession" and grazing on copper toxicity in aquatic microcosms, *Verh. Internat. Verein. Limnol.* 24: 2205-2214.

Thiébaud H., Merlin G., Capovilla M.P., Blake G. (1994). Fate of a volatile chlorinated solvent in indoor aquatic microcosms: sublethal and static exposure to 14C-dichloromethane, *Ecotoxicol. Environ. Safety* 28: 71-81.

Tidou A.S., Moreteau J.C., Ramade F. (1992). Effects of lindane and deltamethrin on zooplankton communities of experimental ponds, *Hydrobiologia* 232: 157-168.

Triffault-Bouchet G. (2004). Effets sur les écosystèmes aquatiques lentiques des émissions de polluants provenant de différents modes de valorisation/élimination de déchets. Application à des mâchefers d'incinération d'ordures ménagères et à des boues de dragage de canaux, Thèse de Doctorat de l'Université de Savoie (spécialité: Biologie et Biochimie Appliquées), 311 pages.

Triffault-Bouchet G., Clément B., Blake G. (2005a). Ecotoxicological assessment of pollutant flux released from bottom ash re-used in road construction, *Aquatic Ecosystem Health Management* 8: 405-414.

Triffault-Bouchet G., Clément B., Blake G. (2005b). Assessment of contaminated sediments with an indoor freshwater/sediment microcosm assay, *Environ. Toxicol. Chem.* 24: 2243-2253.

Van Donk E., Prins H., Voogd H.M., Crum S.J.H., Brock T.C.M. (1995). Effects of nutrient loading and insecticide application on the ecology of *Elodea*-dominated freshwater microcosms. I. Responses of plankton and zoopanktivorous insects, *Arch. Hydrobiol.* 133: 417-439.

Velluet C. (2009). Modélisation de la dynamique algale dans un microcosme aquatique de laboratoire avec renouvellement d'eau, rapport de Travail de Fin d'Etudes de l'ENTPE, 76 pages.

Verrhiest, G., Clément, B. & Blake, G. (2001). Single and combined effects of sediment-associated PAHs on three species of freshwater macroinvertebrates, *Ecotoxicology* 10: 363-372.

Verrhiest G., Clément B. & Merlin G. (2000). Influence of sediment organic matter and fluoranthene-spiked sediments on some bacterial parameters in laboratory freshwater / formulated sediment microcosms, *Aquat. Ecosystem Health Manag.* 3: 359-368.

Verrhiest G., Clément B., Volat B., Montuelle B., Perrodin Y. (2002a). Interactions between a polycyclic aromatic hydrocarbons mixture and the microbial communities in a natural freshwater sediment. *Chemosphere* 46: 187-196.

Verrhiest G., Cortes S., Clément B., Montuelle B. (2002b). Chemical and bacterial changes during laboratory conditioning of formulated and natural sediments, *Chemosphere* 46: 961-974.

Yosowidagdo R. (1996). Évaluation écotoxicologique en microcosmes aquatiques de laboratoire: mise au point méthodologique. Thèse de l'Université de Savoie, Chambéry.

In: Microcosms
Editor: Christopher C. Harris
ISBN: 978-1-62618-661-3

Chapter 4

ASSESSMENT ON TRANSFORMATION OF ORGANIC POLLUTANTS IN MICROCOSMS: A REVIEW

***Xuewei Peng*[1,2] *and Xianguo Li*[1*]**

[1]Key Laboratory of Marine Chemistry Theory and Technology, Ministry of Education, Ocean University of China, Qingdao, China

[2]College of Chemistry and Chemical Engineering, Weifang University, Weifang, China

ABSTRACT

Investigation on transformation of organic pollutants in microcosm is an important way of understanding their fate and behavior in the environment. The common methods used in microcosm experiment include (1) determination of concentration change of organic pollutants during the whole process by gas chromatography (GC) or high performance liquid chromatography (HPLC); (2) identification of degradation products by gas chromatography-mass spectrometry (GC-MS) or high performance liquid chromatography-mass spectrometry (HPLC-MS).

Over the past decades, a new and innovative method of compound-specific isotope analysis (CSIA) using gas chromatography coupled isotope ratio mass spectrometry via a combustion interface (GC-C-IRMS) has been established as a powerful tool for assessing transformation of organic pollutants based on the fractionation of stable isotopes during degradation process.

Two stable isotopes, ^{13}C and ^{2}H, have been extensively exploited to evaluate the transformation processes of contaminants. The enrichment of $\delta^{13}C$ and $\delta^{2}H$ offer a direct evidence of transformation, which does not prominently influenced by physical processes such as volatilization and sorption. Moreover, isotopic fractionation effect is closely associated with the reaction mechanism and the type of bond being broken. In this chapter, we review studies on degradation pathway of organic pollutants in microcosm by mean of CSIA. Especially with diethyl phthalate (DEP) as an example, we discuss how the CSIA technique combined with GC-MS analysis was used for studying the degradation pathway of organic pollutants in microcosm.

* Corresponding author: Xianguo Li, Tel: +86-532-66781761; Fax: +86-532-66782540, E-mail address: lixgouc@126.com (X. Li).

1. INTRODUCTION

With the rapid development of industry and agriculture, more and more toxic chemicals have been released to environment through application, transportation, and accident, leading to environmental deterioration. There is continuing concern about the removal of organic contaminants in natural environment. To date, a variety of remediation strategies have been built to remove these contaminants from the environment. Previous studies showed that biotic and abiotic degradation in the environment is the only process that made organic contaminants nontoxic [1-2]. Full understanding the transformation processes of organic contaminants between and within environmental compartments is important to accurately design and calibrate models to predict the fate of organic contaminants in the environmental systems.

Microcosms are miniature constructed ecosystems in which physical and biological constraints are imposed to enable the controlled study of ecological processes. Owing to a reduction in size or complexity, microcosm may be created artificially or they may represent portions of natural ecosystems, and can be brought into the laboratory for study. Application of microcosms in the study on environmental contaminants has been criticized since substantial information on the behavior of chemicals in the environment can be provided by this approach [3], which are relevant to actual environmental occurrences, and be of potential help for understanding of the concerning **environment issues**.

2. COMMON METHOD FOR ASSESSING TRANSFORMATION OF ORGANIC POLLUTANT

To know the fate and behavior of contaminants in the environment, studies on transformation processes of them have been widely done, which provide valuable information for the design, optimization and risk assessment of decontamination treatments. To date, the evaluation of transformation of organic pollutants in microcosm study has usually been done by means of determination concentrations of parent compounds [4-8], identification of intermediates [9-10], and determination the production of end-product (e.g., CO_2) [9, 11]. Among these methods, the most common method is determination concentrations of parent compound by GC or HPLC, and the evaluation of transformation processes is easily done based on the variations in concentration of contaminant . For example, a variety of volatile organic compounds (VOC), such as monoaromatic hydrocarbons and chlorinated solvents, are typical sediment and groundwater pollutants. Attenuation of these compounds is usually evaluated by GC analysis. Based on the variations in concentration of chloroaliphatic hydrocarbons and methane determined with GC coupled electron capture detector (ECD), Leahy and Shreve studied the effect of organic carbon content on reductive dehalogenation of tetrachloroethylene (PCE) in anaerobic microcosms constructed using leachates, and found the extent of dehalogenation was correlated with the availability or quality of organic carbon in the leachates [4]. Bjerg et al. investigated the degradation of seven aromatic and four chlorinated aliphatic compounds in laboratory batch experiments and in situ microcosm based on the concentration of target compounds analyzed by GC-FID and GC-ECD [5]. The results showed that laboratory batch experiments might be a useful tool in determining the degradability of monoaromatic hydrocarbons and chlorinated aliphatic hydrocarbons. Besides VOCs in the environment, a number of polar compounds are also present in the environment, which are usually detected by HPLC, another instrument commonly used to detect the concentration of organic compounds. For example, to understand the transformation of agrochemicals in the environment and degradation of their intermediates, Lin et al. studied the degradation of three herbicides (atrazine, cyanazine and dicamba) in laboratory microcosms under simulated methanogenic conditions [8]. Samples were analyzed with HPLC, and the results showed that atrazine was much more resistant to degradation than cyanazine and dicamba over 300 days of incubation.

During the course of transformation, some intermediates formed by parent compound will not be decomposed in time due to the adaption of microorganism to these products, and accumulated in the tested system. Compared with blank experiment, identification of these compounds can be achieved by GC-MS (or HPLC-MS) analysis, which will attribute to better understand the degradation processes of contaminants. Combined intermediates information and concentration analysis, Arias et al. assessed **polycyclic aromatic hydrocarbons** (PAH) degradation and metabolite formation in soils microcosm [9], and found that in inoculated microcosms 35% of the added phenanthrene was depleted, 19% being recovered as CO_2, and 3% as diphenic acid.

In addition to methods mentioned above, determination the production of end-product (e.g., CO_2) formed during transformation processes provides an indirect evidence of degradation, which is usually measured by GC-TCD in a closed system. Also, the quantity of total CO_2 produced over the course of biodegradation can be determined by acid–base titration [9].

The common methods are simple and effective in illustrating the transformation of organic contaminants in the environment. Some shortcomings, however, are existed in these methods: (1) besides the biodegradation processes, physical processes such as evaporation, dispersion and sorption, might cause the reduction of the concentration, where no transformation occurs in these processes; (2) owing to complexity of environment system, substances with the same chemical composition or distribution characteristics might be derived from the degradation of different substances or the interaction of the different sources; (3) most metabolites, which are more susceptible to degrade than parent compounds, resulting in very low concentrations, are difficult to be detected; (4) the production of CO_2 was not necessarily resulted from the transformation of target compound since some microorganism might also use other organic compounds in the environment as carbon and energy source, leading to the high estimation to transformation of contaminants.

Therefore, these traditional methods are subjected to challenge in assessing the transformation of organic contaminants in the environment. To accurately assessing transformation of contaminants, new methods are required.

3. Application of Stable Isotope Signature During Transformation of Contaminant

Over the past decades, CSIA, a new technique using gas chromatography coupled isotope ratio mass spectrometry via a combustion interface (GC-C-IRMS), has been developed into a mature analytical method and successfully employed to evaluate the transformation of organic contaminants in the environment [12-15].

3.1. Basic Concept

During chemical and biochemical reaction, activation energies needed to break or form chemical bonds formed by light isotope (e.g., ^{12}C) is lower than that formed by heavy isotope (e.g., ^{13}C), and therefore light isotopes form weaker bonds and react faster than the heavy isotopes, leading to an intrinsic kinetic isotope effect (KIE), which can be defined as

$$\mathrm{KIE} = {}^{12}k/{}^{13}k \tag{1}$$

where ^{12}k and ^{13}k are the specific reaction rates of the light and heavy atom of the element C, respectively.

Owing to the existence of KIE, normal isotopic fractionation usually occurs as the reaction proceeds, leading to the heavy isotope becoming increasingly enriched in the residual substrate. With CSIA, the relative abundance of the heavy and light isotopes in a molecule of a given compound is determined, expressed by the ratio $\delta^{13}C$,

$$\delta^{13}C = \frac{R_{\mathrm{sample}} - R_{\mathrm{standard}}}{R_{\mathrm{standard}}} \cdot 1000\ ‰ \tag{2}$$

$$R = {}^{13}C / {}^{12}C \tag{3}$$

where R_{sample} is the average isotope ratio of element C in a given sample, and R_{standard} is the isotope ratio in a standard reference material.

The trend of isotope enrichment in the remaining substrate can be described by the Rayleigh equation for an element E (e.g., C) in a given molecule (eq. 4) [16].

$$\ln\left(\frac{1000 + \delta^{13}C}{1000 + \delta^{13}C_0}\right) = (\alpha - 1).\ln f \tag{4}$$

$$\varepsilon_{bulk} = 1000(\alpha - 1) \tag{5}$$

where $\delta^{13}C$ is the isotopic composition of the substrate at a given time, $\delta^{13}C_0$ is the initial isotopic composition of the substrate, α is the isotopic fractionation factor, and $f = c_t/c_0$ is the fraction of the remaining substrate. The value of α is obtained by plotting ln$f$ versus $\ln(1000+\delta^{13}C/1000+\delta^{13}C_0)$ and determining the slope of the linear regression. Bulk enrichment factor ($\varepsilon_{\mathrm{bulk}}$) is obtained according to equation 5.

Based on enrichment factor obtained in laboratory culture experiments, the extent of degradation of contaminants in an open microcosm system, such as wetland microcosm, in which material balance might be not accurately built, can be estimated using equation 6.

$$B = 1 - f = 1 - \left(\frac{1000 + \delta^{13}C}{1000 + \delta^{13}C_0} \right)^{1000/\varepsilon} \tag{6}$$

3.2. Evaluation Transformation with Isotope Enrichment

During the course of degradation, light isotopes are transferred preferentially from the reactant to the product pool (normal isotope fractionation) due to KIE, resulting in an enrichment of heavy isotope in the residual substrate. CSIA can provide the unique possibility of identifying transformation processes in the environment based on isotope enrichment of given element in the residual substrate since physical processes such as sorption, vaporization and migrate can not result in observed isotope enrichment [17-20]. In the past decades, studies of CSIA in environment science focused mainly on chlorinated hydrocarbons, methyl *t*-butyl ether (MTBE), and volatile petroleum hydrocarbons due to the following reasons (1) the widespread occurrence of them in the environment; (2) they can be extracted easily from water for GC-C-IRMS analysis.

3.2.1. Evaluation Transformation by Stable Isotope Fractionation in the Residual Substrate

(1) Single-Element Isotopic Fractionation

As one of the basic elements constituting organic compound, carbon has two stable isotopes, with abundance of 98.89% and 1.11% for ^{12}C and ^{13}C, respectively [21]. **Owing to the low** natural **abundance ^{13}C,** generally only quite small percentage of ^{13}C atom will be in the molecule. When ^{13}C atom is present in the reaction center, ^{13}C isotope enrichment in the residual reactants will occur after degradation. The $\delta^{13}C$ values measured by GC-C-IRMS can therefore be used to confirm the degradation of organic contaminants.

As a class of typical contaminants in groundwater, *stable carbon isotopic fractionation during degradation of chlorinated solvents* has been investigated with a variety of microcosms [15, 22-26]. Significant ^{13}C enrichment occurred during anaerobic biodegradation of trichloroethylene (TCE), with a shift of $\delta^{13}C$ more than 15‰ after 90% TCE degradation, and the isotopic fractionation process was described well by a fractionation factor (α=0.9929) [15]. The large enrichment in $\delta^{13}C$ value was similar to that of microbial dechlorination of PCE to ethene in a laboratory microcosm [22], in which all dechlorination steps were accompanied by stable carbon isotopic fractionation, with a shift of 24.2‰ and 34.5‰ for dechlorination of cis-1,2-dichloroethene and vinyl chloride, respectively

In addition to chlorinated solvents, another important contaminant in groundwater is MTBE, a fuel oxygenate used in gasoline to diminish carbon monoxide emissions and enhance octane, which has also been studied by many group with CSIA in recent years[27-33]. Hunkeler et al. [28] reported ^{13}C enrichment of 5.1 to 6.9‰ (f<0.05), and ε_c ranging from -1.5 to -2.0‰ during aerobic biodegradation of MTBE. Lessser et al. [29] observed similar results for aerobic degradation of MTBE in a laboratory microcosm, with a $\delta^{13}C$ shift of 5.5 to 6.4‰ and ε_c of -1.4‰, respectively. Difference with the small ^{13}C enrichment during aerobic biodegradation of MTBE, a large enrichment in $\delta^{13}C$ value was often observed under anaerobic conditions. Kolhathar et al. found much larger ^{13}C enrichment ($\Delta\delta^{13}C$=33.4‰) during anaerobic degradation of MTBE in laboratory microcosm prepared with groundwater and sediment [33]. The estimated ε_c was -9.16‰, which was also considerably larger than that reported during aerobic degradation. The higher ε_c and $\Delta\delta^{13}C$ values observed under anaerobic conditions imply that isotopic fractionation might be related to the specific mechanisms between anaerobic and aerobic MTBE biodegradation.

In complex environment system, the magnitude of isotopic fractionation of a given element can rely on many factors. In some experiments, the transformation of contaminant occurred based on the variations in concentration of contaminant, but no measurable ^{13}C enrichment is present. Lollar et al. investigated aerobic degradation of toluene using laboratory microcosm [15], and found no change in $\delta^{13}C$ value of the residual toluene was observed within analytical uncertainty (± 0.5‰) after degradation. Causes for this might be related to the biodegradation mechanism, in which the bond conversion is often preceded by formation of enzyme-substrate complexes. The reverse step of this preceding process is very slow, making no or only minor isotopic discrimination be observed in the remaining substrate [16]. In contrast, in another study of degradation of toluene and *o*-xylene in anoxic laboratory soil percolation columns [34], a significant increase in $\delta^{13}C$ values ($\Delta\delta^{13}C$>20‰) with decreasing of toluene and *o*-xylene concentration was observed. Under oxic and anoxic conditions, the trend of isotopic fractionation during toluene degradation was consistent with that of chlorinated solvents under both conditions, indicating these contaminants are more easily biodegraded under anoxic conditions.

In addition to degradation conditions (such as oxic and anoxic conditions), the degree of ^{13}C isotopic fractionation of given compound is also influenced by the number of carbon atoms in a given molecule [35]. Causes for this was mainly due to the dilution effect existed in the intrinsic isotopic fractionation, and the higher the number of respective atoms per molecule, the more diluted becomes the extent of analyzed isotopic fractionation. Compared ^{13}C enrichment in compounds of TCE (or PCE) (2 carbon atoms), MTBE (5 carbon atoms), and aromatic hydrocarbon (>6 carbon atoms) mentioned above, one can conclude that the extent of isotope enrichment becomes small with the number of carbon atoms increasing in a given compound. This conclusion was also confirmed by our recent study on aerobic degradation phhalic acid esters (PAEs) (dimethyl phthalate (DMP), di-*n*-butyl phthalate (DBP) and dioctyl phthalate (DOP)) with marine sediments and sea water microcosm [36], in which a small enrichment in $\delta^{13}C$ values was observed after degradation, with $\delta^{13}C_{DBP}$=1.92±0.23‰ (*f*=0.16) and $\delta^{13}C_{DOP}$=0.55±0.21‰ (*f*=0.08). In contrast, a slight large enrichment in $\delta^{13}C$ values ($\delta^{13}C_{DMP}$=+2.05±0.21‰) occurred after 83 % DMP of degradation.

Besides biotic degradation, abotic degradation process also can be evaluated using CSIA. In our previous study, we investigated the behavior of stable carbon isotope during photochemical degradation of DEP in nature sea water under UV conditions. The results showed that an obvious ^{13}C enrichment was observed after 43% photolysis of DEP, with a $\delta^{13}C$ shift of 2.92‰. After more than 90% of DEP converted, a profound ^{13}C isotope enrichment ($\Delta\delta^{13}C$=12.3±0.3‰) occurred, which was clearly an indicator of its photolysis [37]. The large difference of ^{13}C isotopic fractionation in the residual PAEs (such as in DMP and DEP) between biotic and abiotic degradation was mainly related to different degradation pathways during degradation processes. In biodegradation, the bond conversion is often preceded by formation of enzyme-substrate complexes, leading to no or only minor isotopic discrimination in the remaining substrate. During direct photochemical degradation, however, no slow or non-fractionation process (e.g., adsorption to reactive surfaces or formation enzyme-substrate complexes) exists, which is contributed to ^{13}C-enrichment in the residual molecule.

(2) Two-Dimension Element Isotopic Fractionation

During chemical and biochemical reaction, if the two bonded atoms belong to different elements and each of them has stable isotope (such as C, H, O, N, and Cl), isotopic fractionation might be present in the two elements. For example, in addition to carbon, hydrogen is another basic element constituting organic compound, which has two stable isotopes (^{1}H and ^{2}H) in nature. When C−H bond is the reaction center, an observed carbon and hydrogen isotopic fractionation would occur during the course of degradation. This expectation has been confirmed by a number of studies [30-31, 38]. With a microcosm constructed with sediment and groundwater from an MTBE-contaminated aquifer located at VAFB Site, Gray et al. investigated the aerobic degradation of MTBE [31], and found that only small carbon isotope enrichment factors were present during transformation processes, with range from -1.5±0.1‰ to -1.8±0.1‰ and -2.0±0.1‰ to -2.4±0.3‰ for a microcosm from a different contaminated site and strain PM1, respectively. In contrast, significant hydrogen isotope enrichment factors, with range between -29±4‰ and -66±3‰, were obtained in microcosm duplicates. Whereas hydrogen isotopic enrichment factor for MTBE degradation by strain PM1 was -33±5‰ to -37±4‰, which was in a narrow range. The large variations in isotopic fractionation between the two types of microcosm indicated the higher level of complexity in such systems compared with pure strain studies and the sensitivity of hydrogen isotopic fractionation.

In fact, the minor variation of environment conditions, such as redox, the type of microbial and electronic acceptor, will significantly affect the degradation of contaminant in the environment (including degradation rate and degradation pathway) and the isotopic fractionation of given element. To investigate the effect of environment conditions on transformation, Mancini et al. [38] detected the carbon and hydrogen isotopic fractionation during biodegradation of benzene under nitrate-reducing, sulfate-reducing, and methanogenic conditions. After degradation, no significant difference among the $\delta^{13}C$ values of benzene obtained under the three conditions, which was 7.5‰ (f =0.04), 7.1‰ (f =0.08), and 8.2‰ (f =0.08), respectively. In contrast, substantial hydrogen isotopic fractionation in the remaining benzene occurred, and the absolute enrichment in $\delta^{2}H$ under three conditions were 39‰ (f<0.2), 106‰ (f<0.3), and 146‰ (f<0.05), respectively. The large difference in $\delta^{2}H$ enrichment during degradation of benzene under three conditions indicated that the degradation of organic pollutant was closely associated with the redox conditions. Combined the results of hydrogen isotopic fractionation reported by Mancini et al. [38] and Gray et al. [31], one can conclude that the degree of hydrogen isotopic fractionation is very sensitive to the specific environment condition. More over, compared the degree of isotopic fractionation of stable carbon and hydrogen in a same compound, one can conclude that stable hydrogen isotopic fractionation is much larger than that of stable carbon during degradation. The major cause for this is that KIE is associated with the mass of the bonded atom, and the larger the difference in mass between two stable isotopes of given element, the larger the KIE is. Of all the elements, stable hydrogen isotope, ^{1}H and ^{2}H, has the largest difference in mass. Therefore, a large hydrogen isotope enrichment might be expected when the reaction bond including hydrogen. The large difference in the results of isotopic fractionation between stable carbon and hydrogen during degradation imply that stable hydrogen isotopic fractionation is more sensitive than that of stable carbon isotopic fractionation. Based on this concept, stable carbon isotope analysis is more suitable for determining the origin of organic pollutant in contaminated environment, especially with compounds containing high number of carbon atom; whereas stable hydrogen isotope analysis is more adapted to provide evidence of organic pollutants transformation. In sum, the application of two-dimension (2-D) CSIA in the study on transformation of contaminant is more reliable than that of single-element isotope analysis.

3.2.2. Evaluation Transformation by ^{13}C Incorporation to Biomass

During biodegradation, microbial capable of degradation contaminant will incorporate carbon derived from the substrate into their biomass (e.g., fatty acids (FA) or nucleic acids (NA)) [39-44]. The decrease in concentration of contaminant can be associated with an increase in microbial biomass when it is used as a substrate for microbial growth. If the substrate is labeled with ^{13}C, ^{13}C isotope will enriched in biomass after degradation, and the transformation of substrate can be confirmed according to the ^{13}C-label recovered from the biomass. Using a laboratory microcosm prepared with sediments from contaminant aquifer, Pelz et al. [40] investigated the degradation of [methyl-^{13}C]toluene under anoxic denitrifying conditions. Based on ^{13}C-enriched in bacterial phospholipid fatty acids (PLFA) after depletion of toluene, the origin of carbon from [methyl-^{13}C]toluene relative to that from indigenous organic substrates was calculated, which was 45% in 16:1ω7c, 23% in 16:0, and 27% in 18:1ω7c from contaminated sediment, and 80% in 16:1ω7c, 39% in 16:0, and 20% in 18:1ω7c from pristine sediment, respectively.

Culture-dependent laboratory microcosm study, which is often used to investigate the potential of in situ biodegradation, might not reliable since the majority of microbial have not been cultured in this system. Moreover, the composition of laboratory microcosm is certainly different to the actual natural environment, complicating the interpretation of the experiment results. Thus, limitations are present in these laboratory approaches, particularly in providing reliable data to transfer the results to the actual on going in situ processes. To overcome the limitations of ex situ laboratory test systems, an in situ microcosm system (BACTRAP®) has been developed in recent years, which consists of pellets containing activated carbon that can be loaded with isotope (e.g., ^{13}C) labelled substrate and is directly placed in the contaminated environment to study in situ biodegradation of organic pollutants. Owing to the worldwide occurrence in groundwater, the degradation of monochlorobenzene (MCB) has gained more attention. In situ microcosm loaded with ^{13}C labelled MCB has been employed to investigate the biotransformation of MCB in the environment. Braeckevelt et al.[41] investigated in situ biodegradation of MCB and found that the significant isotope composition of FA ($\delta^{13}C_{FA}$) extracted from the BACTRAPs amended with [$^{13}C_6$]-MCB was observed, with range between -39‰ and 7244‰, which was significantly higher than the typical natural abundance of $\delta^{13}C_{FA}$ (<-20‰) found in soil and aquifer material [45-46]. Owing to the enrichment of ^{13}C isotope in FA (>0‰) can only originate from the microbial utilisation of [$^{13}C_6$]-MCB as a carbon source, therefore the level of incorporation of ^{13}C into bacterial biomass is a direct indicator of in situ degradation of MCB.

Although the information on biodegradation provided by in situ microcosm is more close to the actual system, important information on degradation products might be difficult to be obtained owing to the low concentration of contaminant and the open system. Full understanding the transportation of organic pollutants in the environment, combination of in and ex situ microcosm in the study on biodegradation is required, which can provide more information on transportation of contaminants. Nijenhuis et al. [42] investigated MCB degradation with in situ microcosm and laboratory microcosm. Significant ^{13}C isotope enrichment in FA from [$^{13}C_6$]-MCB amended in situ microcosms were observed at several of the investigated depths (m2, m3, m5, and m6), with an isotope signature of $\delta^{13}C>1000‰$, indicating the incorporation of MCB-derived carbon into biomass. More over, ^{13}C labeled benzene was observed in laboratory microcosm. The results in laboratory and in situ microcosms provided clear evidence for biodegradation of MCB, indicating that the use of ^{13}C labeled tracer in combination with in situ and laboratory microcosms was very valuable for analysis of the slow degradation of recalcitrant organic chemicals.

Compared with microcosm added single-substrate, the system of multi-substrate coexisted is more meaningful to evaluate the remediation strategy, which is more close to the actual system and offers reliable information of environment. Stelzer et al. investigated in situ degradation of benzene and toluene with different geochemical zones of contaminated aquifers [43]. The composition of stable carbon isotope of the FA from the benzene and toluene showed that the transformation of contaminant derived carbon into FA was higher in toluene compared to benzene amended BACTRAPs, indicating that toluene is more degradable than benzene under anaerobic conditions, which is similar to that reported by Geyer et al. [44], who also found that the enrichment of ^{13}C in FA from toluene is much higher than that from benzene.

In addition to evaluation the transformation of contaminant, the activity of microorganism during the course of degradation can be revealed based on the extent of ^{13}C incorporation into specific FA, which is not reviewed in this chapter. In summary, in situ microcosms approach provide a very promising tool to directly test in situ biodegradation of contaminants within reasonable time, which opens prospects for detailed analysis of the microbial key-players in future.

3.2.3. Evaluation Transformation Through Degradation-Product

During the course of transformation, daughter compound formed by parent compound might not be degraded in time, and accumulated in the environment. There is a relationship between the primary product and the precursor in composition of stable isotope. The isotope composition of daughter compound may also be used to identify the occurrence of transformation. In particular the substrate employed in the test system is a ^{13}C-labelled compound, the $\delta^{13}C$ values of the primary daughter compound are more easily to offer an evidence of transformation based on the tracer of ^{13}C-label. During degradation [$^{13}C_6$]-MCB, Nijenhuis et al. [42] observed the occurrence of significantly higher ^{13}C enrichment of [$^{13}C_6$]-benzene ($\delta^{13}C>1000‰$) on the BACTRAPs at different depth of in situ microcosm. These results imply that [$^{13}C_6$]-benzene might be a possible intermediate, and demonstrated the occurrence of reductive dehalogenation of MCB. This conclusion was consistent with that reported by Braeckevelt et al., who investigated the degradation of MCB with in situ microcosm [41], and also found significantly higher amounts of [$^{13}C_6$]-benzene on the BACTRAPs from both planted and unplanted segment.

Although the evidence of transformation can be provided by determination isotope composition of the degradation product, the extent of fractionation of degradation product, however, can not be described with Rayleigh equation, and therefore the fractionation factor can not be obtained [24]. The reason is that the application of Rayleigh equation is depended on the relation between the isotopic enrichment values and the strict concentration of substrate. Parent compound can meet this requirement and thus their degradation can be evaluated by Rayleigh equation. As for the degradation product, however, the concentration is affected by the processes of depletion and production [24], and therefore the relation between isotope enrichment value and concentration can not be built accurately. In spite of this, the isotope signature of intermediates also offers information on transformation.

With the time of biodegradation, intermediates formed by parent compound are gradually available to microorganism capable of degradation it. These compounds therefore can be further degraded, and eventually mineralized to CO_2 and H_2O. The determination of the production of CO_2 also provides an evidence of organic contaminants transformation. This method is effective in assessing the transformation process of pollutants in simple system (especially single-substrate as the carbon and energy source). Most study systems, however, are complexity and a variety of substrates coexisted in it, which may be available to microbial action, resulting in an uncertain in determination content of CO_2 with traditional method.

In recent years, ^{13}C-labeled substrate confirmatory determination of $^{13}C\text{-}CO_2$ has been applied to assessing the degradation of contaminants through microcosm study [40, 42, 47]. Pelz et al. investigated the degradation of [methyl-^{13}C]toluene in laboratory microcosm [40]. Based on the $\delta^{13}C$ values of CO_2 from microcosms, controls, and pure cultures, the amount of CO_2 resulted from the degradation of ^{13}C-labeled substrate was calculated, with 11±2% and 54±1% of the total CO_2 originated from mineralization of [methyl-^{13}C]toluene in microcosms with contaminated and pristine sediments, respectively. This result indicated that the specific information on mineralization of ^{13}C labeled substrate was found to be an effective marker to test the fate of one particular contaminant.

In addition to confirm the transformation of contaminants, the determination of $^{13}C\text{-}CO_2$ was also applied to discriminate the degradation of different compound. Morasch et al. [47] investigated the mineralization of ^{13}C-labeled and non-labeled contaminants (benzene, naphthalene, and acenaphthene) in the presence of other unknown organic substances of the sediment material. Under oxic conditions, the maximum ^{13}C isotope enrichment in $^{13}C\text{-}CO_2$ observed in the microcosm added ^{13}*C-la*beled benzene and naphthalene at day 2 of incubation was 411 to 667‰ and 556 to 902‰, corresponding to 0.46% to 0.75% and 0.62% to 1.01% of $^{13}C\text{-}CO_2$ per total CO_2, respectively. In contrast, the maximum ^{13}C isotope enrichment in the microcosm loaded with ^{13}C-labeled acenaphthene observed between day 40 and 135 was 106 to 281‰, corresponding to 0.12% to 0.3% of $^{13}C\text{-}CO_2$ per total CO_2. The degree of ^{13}C enrichment in $^{13}C\text{-}CO_2$ and the appearance time of the maximum $^{13}C\text{-}CO_2$ revealed that compound containing low number of carbon atoms was more available to microbial than that containing high number of carbon atoms, which was consistent with the conclusion our early reported [36]. Under anoxic conditions, the maximum ^{13}C isotope enrichment was occurred at day 9 for microcosm added ^{13}C-labeled benzene, and at day 136 for microcosm added ^{13}C-labeled naphthalene and acenaphthene. The degradation of the three contaminant under oxic conditions was faster than that under anoxic conditions, which was contrary to that of chlorinated solvents and MTBE [30,31], indicating that the degradation of different contaminants might be related to the specific environment conditions or the specific degradation pathway.

3.3. Degradation Pathway of Organic Pollutants with CSIA

3.3.1. Identification of Degradation Pathway by ε (KIE)

Investigations on degradation mechanism of organic pollutants are contributed to better understand the environmental fate of pollutants and to improve environmental remediation strategies [48]. Generally, the degradation pathway of contaminants is studied by identification of degradation products using GC-MS or HPLC-MS, which has an uncertainty [49, 50]. During the course of degradation, isotopic fractionation process can be described with Rayleigh equation. The resulted ε (or $KIE=\alpha^{-1}$) is often strongly relying on the reaction mechanism since they depend on the first irreversible transformation step, typically the first bond breakage. Compared ε (or KIE) with that of known degradation pathways reported in literature, the degradation pathway of contaminates in the environment therefore can be deduced. At early stage of CSIA application, ε was used to speculate the degradation pathway of organic pollutants in the environment. Kuder et al. investigated stable carbon and hydrogen isotope fractionation during anaerobic biodegradation of MTBE using sediment and groundwater microcosm [30]. In their study, the ε_C value of the enrichment cultures (-12.9±1.1‰) is in good agreement with the value of that expected for cleavage of a O–C bond (-12.2‰) [51], and is higher than isotope effects expected from cleavage of C–H bonds. These results suggested that enzymatic hydrolysis of a single $O–C_{methyl}$ bond is the first reaction, which was further confirmed by the intermediate analysis using GC-MS. In another study, a large hydrogen enrichment factors (range of -29‰ to -79‰) was obtained during anaerobic degradation of benzene under nitrate-reducing, sulfate-reducing, and methanogenic conditions [38]. Compared the result with that of aerobic benzene degradation [52], Mancini et al. proposed that the initial step of anaerobic benzene degradation might be cleavage of a C–H bond. Similar to this method, *using KIE*, Hirschorn et al. [25] evaluated the pathway of aerobic biodegradation 1,2-dichloroethane in the East and West Louisiana (EL, WL) microcosm. The resulted KIE (1.01 and 1.06), which was consistent with KIE produced in oxidation (1.00-1.03) [51, 56] and hydrolytic (1.04-1.09) [51, 57] dehalogenation (S_N2) reactions, respectively, indicated the degradation pathway of aerobic biodegradation 1,2-dichloroethane.

3.3.2. Identification of Degradation Pathway by AKIE

The isotope signature measured by GC-C-IRMS is an average isotope composition of the entire reactant molecule rather than the isotope ratios at a specific reaction position. Therefore, the observed bulk isotopic fractionation or ε_{bulk} obtained by eq. 5 often do not directly reflect the underlying isotope-sensitive process, and the information on specific KIE at different reactive positions of reactant molecule can not be obtained, rendering the extraction of degradation pathway more complex. To correctly predict the underlying degradation pathway, ε_{bulk} have to be converted to apparent kinetic isotope effects (AKIE) by Rayleigh equation using isotope data determined in laboratory microcosm studies under defined conditions. Owing to considering the presence of heavy isotopes in non-reactive positions and intermolecular competition between several reactive positions in one molecule, such AKIE values can describe the isotope effect associated with the cleavage of a specific bond and provide insight into the degradation mechanism.

To obtain AKIE, the value of reaction-position isotope enrichment (e.g., $\Delta\delta^{13}C_{rp}$) is calculated based on equation 7. The value of reaction-position isotope enrichment factor (ε_{rp}) is achieved by plotting ln$f$ versus ln(1000+$\delta^{13}C_0$+n/x.$\Delta\delta^{13}C_{bulk}$ /1000+$\delta^{13}C_0$) based on equation 8 and determining the slope of the linear regression.

$$\Delta\delta^{13}C_{rp} = {}^{n}\!/_{x}.\Delta\delta^{13}C_{bulk} \tag{7}$$

$$\ln\frac{R}{R_0} = \ln\frac{(1000+\delta^{13}C_0 + {}^{n}\!/_{x}.\Delta^{13}C_{bulk})}{(1000+\delta^{13}C_0)} = \frac{\varepsilon_{rp}}{1000}\ln f \tag{8}$$

where n is the total number of carbon atoms in the molecule, x is the number of carbon atoms at reaction-position, $\Delta\delta^{13}C_{bulk}$ is the isotope enrichment value in $\delta^{13}C$ from t_0 to t, and $\delta^{13}C_0$ is the initial value at t_0. AKIE can then be calculated with the following equation 9 [16].

$$AKIE = \left(\frac{{}^{l}k}{{}^{h}k}\right)_{apprent} = \frac{1}{\alpha_{rp}} = \frac{1}{1+1000\varepsilon_{rp}} \tag{9}$$

In the case of symmetric molecules, there is not only intermolecular isotopic competition between light (e.g., ^{12}C) and heavy (e.g., ^{13}C) isotope, but also intramolecular competition between the reaction positions. To obtain a correct AKIE, a further correction for intramolecular competition to the molecule is necessary to perform. The AKIE can be obtained according to equation 10 [16].

$$AKIE = \left(\frac{{}^{l}k}{{}^{h}k}\right)_{apprent} = z.\frac{1}{\alpha_{rp}} = z.\frac{1}{1+1000\varepsilon_{rp}} \tag{10}$$

where z, the factor introduced for intramolecular competition, is the number of chemically equivalent reactive positions.

In our previous study, the photochemical degradation of DEP in nature sea water under UV conditions was investigated with CSIA [37]. The $\delta^{13}C$ values in the residual DEP was shown in Fig. 1.

Enrichment factor (ε_{bulk}=-5.95±1.26‰.) is obtained by a linear regression of the isotope data according to equation 4. Early studies on photolysis of PAEs showed that the initial step was hydrolysis of the ester group by cleavage of C–O bond [54-55].

In DEP molecule (12 carbon atoms), only 2 carbon atoms (carbonyl carbon) are present in the reactive positions, and therefore dilution effects from the nonreactive isotopes is existed when ^{13}C is in the other 10 carbon atom positions. In order to obtain ε_{rp}, $\Delta\delta^{13}C_{rp}$ is first obtained based on equation 8. In this case, n=12, x=2, the ε_{rp} obtained by a linear regression of the isotope enrichment data corrected according to equation 7 is -35.25±2.26‰ (Fig. 1).

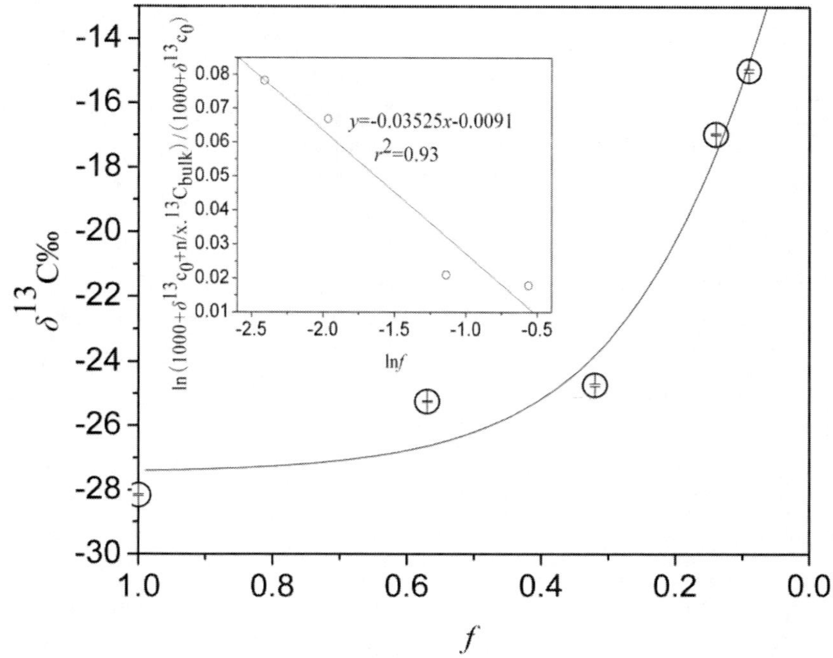

Figure 1. Stable carbon isotope fractionation during photolysis of DEP in seawater [37]. Insert is linear regression of ^{13}C isotope enrichment equation during photolysis of DEP.

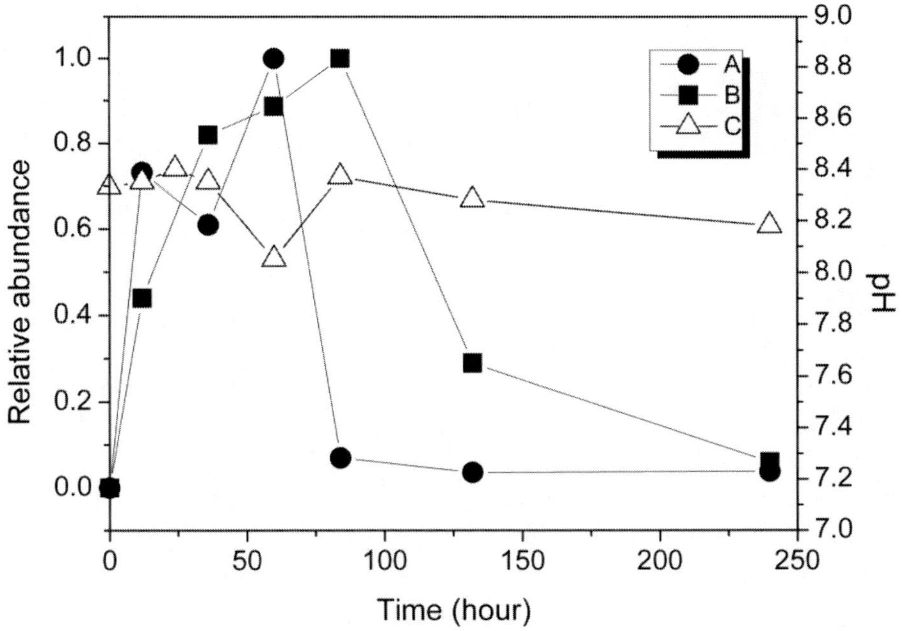

Figure 2. The relative abundance of DEP photolysis products and pH value over time during photolysis at 254 nm. Circles (A) represent MEP, squares (B) represent phthalic anhydride, and triangles (C) represent pH [37].

In addition to considering the dilution effect, in this case, the reaction occurs at one of the two carbonyl carbon atoms, which is chemically equivalent, and there would be intramolecular competition between the two carbonyl carbon atoms within a single DEP molecule. To correct intramolecular competition, a factor of z=2 should be introduced when converting to $AKIE_C$ based on equation 10. The resulted $AKIE_C$ (1.075) is in agreement with that reported in reference (1.048-1.085) [30], indicating that the first step during UV degradation might be cleavage of a C-O bond.

Another case is that the reaction might simultaneously occur at the two carbon atoms based on identification of phthalic acid during degradation DEP [53]. If this assumption is existence, there would be no intramolecular competition within a single DEP molecule, and the corresponding $AKIE_C$ is 1.037, which is profound smaller than semiclassical KIE (1.061) [51]. To further confirm the primary product during degradation, intermediates were identified by GC-MS. Fig.2 shows the relative abundance of DEP photolysis products over time during degradation. Based on intermediate information, the main degradation products were MEP and phthalic anhydride, and the primary intermediate should be MEP in the light of the content over time and the relationship of molecule structure between these two products.

According to the intermediate information and the AKIE analysis, the dominant degradation pathway of DEP in natural seawater is that DEP first transferred to its corresponding excited states, then followed by cleavage of a C–O bond to form MEP. The later further degraded to form phthalic anhydride (Fig. 3). Then ring open occurred, and some intermediates mineralized to CO_2. The results of our study showed that mechanistic information about the initial step of biodegradation can be provided by AKIE.

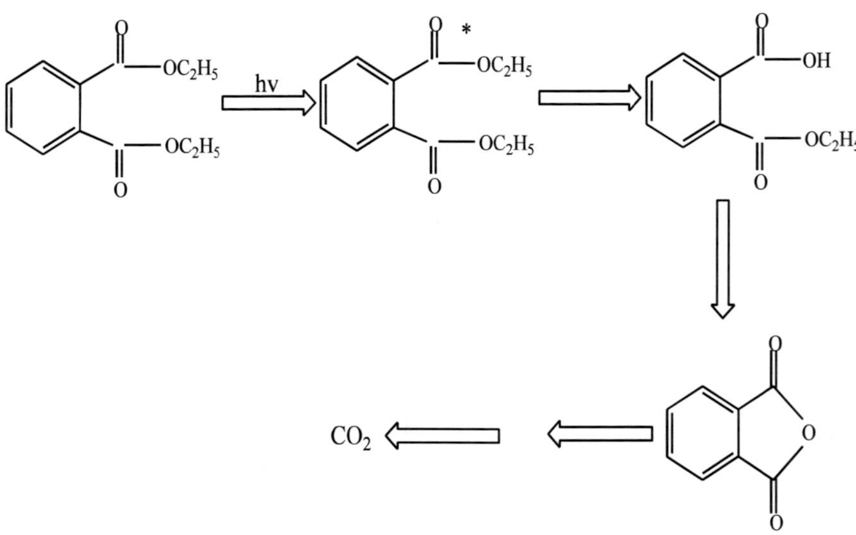

Figure 3. Proposed pathways for DEP photolysis in seawater [37].

3.3.3. Identification of Degradation Pathway with 2-D Isotope Analysis

At early stage of CSIA, the evaluation of degradation mechanism for contaminant was mainly done by isotopic fractionation of single-element (e.g., ^{13}C), which can significantly vary due to the minor variation of experiment conditions or nonisotope-sensitive steps of a particular biodegradation pathway. Therefore, identification of microbial degradation pathways by isotope analysis of only single-element might be a rough estimation, and not always reliable. In addition to isotopic fractionation of given single element (e.g., ^{13}C), the occurrence of isotopic fractionation in the other reaction atom will be present if it has stable isotope. Recently, the correlation (eq.11) of changes in stable isotope ratios for different elements (e.g., C and H) has been developed for characterizing the biodegradation pathways, which is more precisely than by isotope fractionation of only one element. The reason for this is mainly that the effects of nonfractionating processes and masking effects were eliminated since the two bonded atoms are assumed to be affected similarly during the course of degradation.

$$\Lambda = \frac{\Delta\delta^{2}H}{\Delta\delta^{13}C} \approx \frac{\varepsilon H_{\mathrm{bulk}}}{\varepsilon C_{bulk}} \tag{11}$$

Compared with single-element CSIA, the 2-D CSIA approach is relatively straightforward to apply for discriminating different degradation pathways. To investigate the initial reaction step of anaerobic benzene biodegradation, Mancini et al. investigated carbon and hydrogen isotopic fractionation under nitrate-reducing, sulfate-reducing, and methanogenic conditions with laboratory microcosm [58]. The resulted Λ values under sulfate-reducing and methanogenic conditions are higher (29±2) than that under nitrate-reducing condition (13±2). Based on these results, they proposed the mechanism of anaerobic benzene biodegradation: a hydroxylation mechanism with methanogenic cultures and a methylation mechanism with the nitrate-reducing cultures, indicating different mechanisms were present for the same substrate under different conditions. These results demonstrate the availability of Λ in the investigation on degradation mechanism.

In addition to different conditions, the pathway of organic pollutant under the same or similar conditions also gained attention. Fischer et al. investigated stable carbon and hydrogen isotopic fractionation of benzene with systems of laboratory microcosm, column system, and an aquifer transect under similar redox and temperature conditions [59]. A large difference in the values of ε_c and ε_H was present in the experiment of laboratory microcosm and column system, respectively, indicating that ε of single-element might not be used to distinguish benzene degradation. In contrast, the Λ values are similar, with a value of 23±5, 28±3, and 24±2 for laboratory microcosm, column system, and an aquifer, respectively. Based on these observations, they proposed that benzene biodegradation is probably initiated by a fermenting microorganism which lives in syntrophy with sulfate reducers.

A variety of microorganisms exist in the environment system, making the complexity of the study system. At different stages of biodegradation, different degradation pathway might be present for the same substrate. Causes of this might be related to (1) the amount of microbial which controls the primary degradation; (2) available of the substrate to the dominant microbial. Using a freshwater wetland microcosm, Rokoczy et al. investigated biodegradation of benzene with 2-D CSIA [60]. Significantly different Λ values were present at four stages (day 231, 293, 345, 370) of biodegradation. At day 231, stable carbon isotope fractionation was observed, but no hydrogen fractionation occurred. A slope of Λ=1±2 was obtained, suggesting the existence of dihydroxylation pathway. At day 293, a higher Λ value (4±5) was obtained due to the significant fractionation of hydrogen element, indicating the attribution of either dihydroxylation or monohydroxylation, i.e., the existence of a transition phase in which both pathways coexisted. With time of biodegradation, the Λ values of day 345 and day 375 were 9±2 and 9±3, respectively, which means the dominant degradation pathway was monohydroxylation.

However, variable isotopic fractionation factors within the same degradation mechanism might be caused by slight differences in transition states, which are very difficult to determine. This may produce different Λ values even for apparently similar degradation mechanisms. The application of combination of AKIE and Λ would provide more reliable result. To gain deeper insights into benzene biodegradation pathways, the analysis of carbon and hydrogen AKIE values and 2-D isotope fractionation obtained in laboratory microcosms was carried out [61]. The $AKIE_C$ were compared with that reported in literature, which were lower for benzene dihydroxylation than for monohydroxylation and anaerobic degradation, enabling discrimination between these reaction mechanisms. Variations in Λ values confirmed that more than one reaction mechanism might exist for monohydroxylation, and ranges of Λ values were determined for dihydroxylation ($\Lambda<2$), monohydroxylation ($\Lambda=7$–9), andanaerobic degradation ($\Lambda>17$).

3.3.4. Identification of Degradation Pathway by the Tracer of ^{13}C

In addition to methods mentioned above, degradation pathway of contaminants could be obtained by identification of products based on the tracer of stable carbon isotope [28, 41-42]. In particular the substrate, used as a sole carbon and energy source for microbial, was labeled with ^{13}C. Owing to the high abundance of ^{13}C in the target molecule, the abundance of ^{13}C in the primary product formed during degradation is lower than that of the precursor but higher than that existed in nature. The product of labeled substrate therefore can be identified based on the tracer of ^{13}C. In recent years, MCB labeled with ^{13}C has been used as an isotope tracer to evaluate its biodegradation [41-42]. Nijenhuis et al. investigated the anaerobic degradation of MCB using in situ microcosm loaded with $^{13}C_6$-labeled MCB incubated at different depths, and found that ^{13}C-labeled benzene was present at several of the investigated depths with an isotope signature of $\delta^{13}C>1000‰$ [42]. The results of the variations in concentration and the $\delta^{13}C$ signature of benzene showed that MCB is degraded reductively to benzene under anoxic conditions in the aquifer prior mineralization. This result was consistent with the conclusion reported by Bracekevelt et al., who reported that [$^{13}C_6$]-benzene was found on the BACTRAPs during reductive dehalogenation of MCB in a constructed wetland [41]. The pathway reported by Nijenhuis et al. [42] and Bracekevelt et al. [41] would suggest a similar pattern as observed previously by Nowak et al. [62].

CONCLUSION

As an important approach, microcosm has been widely applied to investigate the fate and behavior of organic pollutant in the environment. The evaluation of transformation processes of contaminant in microcosm are mainly done employing traditional methods, and provide valuable information on degradation.

CSIA as an emerging technology has important applications in the investigation of organic pollutants. Stable isotope enrichment of the specific element in the organic contaminants not only can be used as a direct indicator of the biodegradation of pollutants, but also the transformed mechanism can be inferred based on the corresponding ε_E, AKIE, and Λ. The combination of traditional methods and CSIA can make up for the shortcomings of each other, which will provide more reliable information on contaminant degradation.

Although CSIA technology has been successfully used in the investigation of organic pollutants, which focused mainly on volatile and semi-volatile, with respect to a large number of organic pollutants in the environment, CSIA technology is still in its infancy. Better play to the CSIA environmental science, the different chemical bond isotope fractionation database is needed to continue to improve.

ACKNOWLEDGMENTS

This work was financially supported by the National Natural Science Foundation of China (Grant Nos. 40976041 and 20775074).

REFERENCES

[1] Slater, G. F., Lollar, B. S., Sleep, B. E., Edwards, E. A. (2001). Variability in carbon isotopic fractionation during biodegradation of chlorinatedethenes: implications for field applications. *Environ. Sci. Technol.*, 35, 901–907.

[2] USEPA (1991) *"Understanding Bioremediation: A Guidebook for Citizens,"* EPA/540/2-91/002, Office of Research and Development, Washington, D. C.

[3] Suderman, K. and Thistle, D. (2003). A microcosm system for the study of pollution effects in shallow, sandy, subtidal communities. *Environ. Toxicol. Chem.*, 22, 1093-1099.

[4] Leahy, J. G. and Shreve, G. (2000). The effect of organic carbon on the sequential reductive dehalogenation of tetrachloroethylene in landfill leachates, *Water Res.,* 34, 2390–2396.

[5] Bjerg, P. L., Riigge, K., Cortsen, J., Nielsen, P. H., Christensen, T. H. (1999). Degradation of aromatic and chlorinated aliphatic hydrocarbons in the anaerobic part of the Grindsted Landfill leachate plume: In situ microcosm and laboratory batch experiments. *Ground Water*, 37, 113–121.

[6] Lyew, D., Tartakovsky, B., Manuel, M.-F., Guiot, S. R. (2002). A microcosm test for potential mineralization of chlorinated compounds under coupled aerobic/anaerobic conditions. *Chemosphere*, 47, 695–699.

[7] Labana, S., Singh, O. V., Basu, A., Pandey, G., Jain, R. K. (2005). A microcosm study on bioremediation of p-nitrophenol-contaminated soil using arthrobacter protophormiae RKJ100. *Appl. Microbiol. Biotechnol.,* 68, 417–424.

[8] Lin, C., Gu, J. G., Qiao, C., Duan, S., Gu, J. D. (2006). Degradability of atrazine, cyanazine, and dicamba in methanogenic enrichment culture microcosms using sediment from the Pearl River of Southern China. *Biol. Fertil. Soils,* 42, 395–401.

[9] Arias, L., Bauzá, J., Tobella, J., Vila, J., Grifoll, M. (2008). A microcosm system and an analytical protocol to assess PAH degradation and metabolite formation in soils. *Biodegradation*, 19, 425–434.

[10] Meckenstock, R. U., Annweiler, E., Michaelis, W., Richnow, H. H., Schink, B. (2000). Anaerobic naphthalene degradation by a sulfate-reducing enrichment culture. *Appl. Environ. Microbiol.*, 66, 2743–2747.

[11] Li, Y., Yediler, A., Ou, Z., Conrad, I., Kettrup, A. (2001). Effects of a non-ionic surfactant (Tween-80) on the mineralization, metabolism and uptake of phenanthrene in wheat-solution-lava microcosm. *Chemosphere*, 45, 67–75.

[12] Braeckevelt, M., Fischer, A., Kastner, M. (2012). Field applicability of compound-specific isotope analysis (CSIA) for characterization and quantification of in situ contaminant degradation in aquifers. *Appl. Microbiol. Biotechnol.,* 94, 1401–1421.

[13] Meckenstock, R. U., Morasch, B., Griebler, C., Richnow, H. H. (2004). Stable isotope fractionation analysis as a tool to monitor biodegradation in contaminated acquifers. *J. Contam. Hydrol.*, 75, 215– 255.

[14] Morasch, B., Hunkeler, D., Zopfi, J., Temime, B., Höhener, P. (2011). Intrinsic biodegradation potential of aromatic hydrocarbons in an alluvial aquifer-Potentials and limits of signature metabolite analysis and two stable isotope-based techniques. *Water Res.* 45, 4459–4469.

[15] Lollar, B. S., Slater, G. F., Ahad, J., Sleep, B., Spivack, J., Brennan, M., Mackenzie, P. (1999). Contrasting carbon isotope fractionation during biodegradation of trichloroethylene and toluene: implications for intrinsic bioremediation. *Org. Geochem.*, 30, 813–820.

[16] Elsner, M., Zwank, L., Hunkeler, D., Schwarzenbach, R. P. (2005). A new concept linking observable stable isotope fractionation to transformation pathways of organic pollutants. *Environ. Sci. Technol.*, 39, 6896–6916.

[17] Slater, G. F., Ahad, J. M. E., Lollar, B. S., Allen-King, R., Sleep, B. (2000). Carbon isotope effects resulting from equilibrium sorption of dissolved VOCs. *Anal. Chem.*, 72, 5669–5672.

[18] Harrington, R. R., Poulson, S. R., Drever, J. I., Colberg, P. J. S., Kelly, E. F. (1999). Carbon isotope systematics of monoaromatic hydrocarbons :vaporization and adsorption experiments. *Org. Geochem.*, 30, 765–775.

[19] Wang, Y. and Huang, Y. (2003). Hydrogen isotopic fractionation of Petroleum hydrocarbons during vaporization: implications for assessing artificial and natural remediation of petroleum contamination. *Appl. Geo. Chem.*, 18, 1641–1651.

[20] Schüuth, C., Taubald, H., Bolaño, N., Maciejczyk, K. (2003). Carbon and hydrogen isotope effects during sorption of organic contaminants on carbonaceous materials. *J. Contam. Hydrol.*, 64, 269–281.

[21] Faure, G. *Principles of isotope geology*, John Wiley and Sons: New York 1986.

[22] Hunkeler, D., Aravena, R., Butler, B. J. (1999). Monitoring microbial dechlorination of tetrachloroethene (PCE) in groundwater using compound-specific stable carbon isotope ratios: microcosm and field studies. *Environ. Sci. Technol.*, 33, 2733–2738.

[23] Hunkeler, D., Aravena, R., Cox, E. (2002). Carbon isotopes as a tool to evaluate the origin and fate of vinyl chloride: laboratory experiments and modeling of isotope evolution. *Environ. Sci. Technol.*, 36, 3378–3384.

[24] Van Breukelen, B. M., Hunkeler, D., Volkering, F. (2005). Quantification of sequential chlorinated ethene degradation by use of a reactive transport model incorporating isotope fractionation. *Environ. Sci. Technol.*, 39, 4189–4197.

[25] Hirschorn, S. K., Dinglasan, M. J., Elsner, M., Mancini, S. A., Lacrampe-Couloume, G., Edwards, E. A., Lollar, B. S. (2004). Pathway dependant isotopic fractionation during aerobic biodegradation of 1,2-dichloroethane. *Environ. Sci. Technol.*, 38, 4775–4781.

[26] Jeannottat, S. and Hunkeler, D. (2012). Chlorine and carbon isotopes fractionation during volatilization and diffusive transport of trichloroethene in the unsaturated zone. *Environ. Sci. Technol.*, 46, 3169–3176.

[27] Schmidt, T. C., Schirmer, M., Weis, H., Haderlein, S. B. (2004). Microbial degradation of methyl tert-butyl ether and tert-butyl alcohol in the subsurface. *J. Contam. Hydrol.*, 70, 173–203.

[28] Hunkeler, D., Butler, B. J., Aravena, R., Barker, J. F. (2001). Monitoring biodegradation of methyl tert-butyl ether (MTBE) using compound-specific carbon isotope analysis. *Environ. Sci. Technol.*, 35, 676–681.

[29] Lesser, L. E., Johnson, P. C., Aravean, R., Spinnler, G. E., Bruce, C. L., Salanitro, J. P. (2008). An evaluation of compound-specific isotope analyses for assessing the biodegradation of MTBE at port Hueneme, CA. *Environ. Sci. Technol.*, 42, 6637–6643.

[30] Kuder, T., Wilson, J. T., Kaiser, P., Kolhatkar, R., Philp, P., Allen, J. (2005). Enrichment of stable carbon and hydrogen isotopes during anaerobic biodegradation of MTBE: microcosm and fieldevidence. *Environ. Sci. Technol.*, 39, 213–220.

[31] Gray, J. R., Lacrampe-Couloume, G., Gandhi, D., Scow, K. M., Wilson, R. D., Mackay, D. M., Lollar, B. S. (2002). Carbon and hydrogen isotopic fractionation during biodegradation of methyl tert-butyl ether. *Environ. Sci. Technol.*, 36, 1931–1938.

[32] Thornton, S. F., Bottrell, S. H., Spence, K. H., Pickup, R., Spence, M. J., Shah, N., Mallinson, H. E.H., Richnow, H. H. (2011). Assessment of MTBE biodegradation in contaminated groundwater using ^{13}C and ^{14}C analysis: Field and laboratory microcosm studies. *Appl. Geochem.*, 26, 828–837.

[33] Kolhatkar, R., Kuder, T., Philp, P., Allen, J., Wilson, J. T. (2002). Use of compound-specific stable carbon isotope analyses to demonstrate anaerobic biodegradation of MTBE in groundwater at a gasoline release site. *Environ. Sci. Technol.,* 36, 5139–5146.

[34] Richnow, H. H., Annweiler, E., Michaelis, W., Meckenstock, R. U. (2003). Microbial in situ degradation of aromatic hydrocarbons in a contaminated aquifer monitored by carbon isotope fractionation. *J. Contam. Hydrol.*, 65, 101–120.

[35] Morasch, B., Richnow, H. H., Vieth, A., Schink, B., Meckenstock, R. U. (2004). *Appl. Environ. Microbiol.*, 70, 2935–2940.

[36] Peng, X. and Li, X. (2012). Compound-specific isotope analysis for aerobic biodegradation of phthalate acid esters. *Talanta*, 97, 445–449.

[37] Peng, X., Feng, L., Li, X. (2013). Pathway of diethyl phthalate photolysis in sea-water determined by gas chromatography–mass spectrometry and compound-specific isotope analysis. *Chemosphere*, 90, 220–226.

[38] Mancini, S. A., Ulrich, A. C., Lacrampe-Couloume, G., Sleep, B., Edwards, E. A., Lollar, B. S. (2003). Carbon and hydrogen isotopic fractionation during anaerobic biodegradation of benzene. *Appl. Environ. Microbiol.*, 69, 191–198.

[39] Stelzer, N., Imfeld, G., Thullner, M., Lehmann, J., Poser, A., Richnow, H.-H., Nijenhuis, I. (2009). Integrative approach to delineate natural attenuation of chlorinated benzenes in anoxic aquifers. *Environ. Pollut.,* 157, 1800–1806.

[40] Pelz, O., Chatzinotas, A., Andersen, N., Bernasconi, S. M., Hesse, C., Abraham, W.-R., Zeyer, J. (2001). Use of isotopic and molecular techniques to link toluene degradation in denitrifying aquifer microcosms to specific microbial populations. *Arch. Microbiol.*, 175, 270–281.

[41] Braeckevelt, M., Rokadia, H., Imfeld, G., Stelzer, N., Paschke, H., Kuschk, P., Kästner, M., Richnow, H.-H., Weber, S. (2007). Assessment of in situ biodegradation of monochlorobenzene in contaminated groundwater treated in a constructed wetland. *Environ. Pollut.*, 148, 428–437.

[42] Nijenhuis, I., Stelzer, N., Kastner, M., Richnow, H.-H. (2007). Sensitive detection of anaerobic monochlorobenzene degradation using stable isotope tracers. *Environ. Sci. Technol.*, 41, 3836–3842.

[43] Stelzer, N., Büning, C., Pfeifer, F., Dohrmann, A. B., Tebbe, C. C., Nijenhuis, I., Kästner, M., Richnow, H. H. (2006). In situ microcosms to evaluate natural attenuation potentials in contaminated aquifers. *Org. Geochem.*, 37, 1394–1410.

[44] Geyer, R., Peacock, A. D., Miltner, A., Richnow, H.-H., White, D. C., Sublette, K. L., Kastner, M. (2005). In situ assessment of biodegradation potential using biotraps amended with ^{13}C-labeled benzene or toluene. *Environ. Sci. Technol.*, 39, 4983–4989.

[45] Pelz, O., Chatzinotas, A., Zarda-Hess, A., Abraham, W. R., Zeyer, J. (2001). Tracing toluene-assimilating sulfate-reducing bacteria using ^{13}C-incorporation in fatty acids and whole-cell hybridization. *FEMS Microbiol. Ecol.*, 32, 123–131.

[46] Pombo, S. A., Pelz, O., Schroth, M. H., Zeyer, J. (2002). Field-scale ^{13}C-labeling of phospholipid fatty acids (PFLA) and dissolved inorganic carbon: tracing acetate assimilation and mineralization in a petroleum hydrocarbon-contaminated aquifer. *FEMS Microbiol. Ecol.*, 41, 259–267.

[47] Morasch, B., Höhener, P., Hunkeler, D. (2007). Evidence for in situ degradation of mono-and polyaromatic hydrocarbons in alluvial sediments based on microcosm experiments with ^{13}C-labeled contaminants. *Environ. Pollut.*, 148, 739–748.

[48] Pieper, D. H., Santos, V. A., Golyshin, P. N. (2004). Genomic and mechanistic in sights into the biodegradation. *Curr. Opin. Biotechnol.*, 15, 215–224.

[49] Zwank, L., Berg, M., Elsner, M., Schmidt, T. C., Schwarzenbach, R. P., Haderlein S. B. (2005). New evaluation scheme for two-dimensional isotope analysis to decipher biodegradation processes: application to groundwater contamination by MTBE. *Environ. Sci. Technol.*, 39, 1018–1029.

[50] Suflita, J. M. and Mormile, M. R. (1993). Anaerobic biodegradation of known and potential gasoline oxygenates in the terrestrial subsurface.*Environ. Sci. Technol.*, 27, 976–978.

[51] Huskey, W.P. Origins and interpretations of heavy-atom isotope effects. In *Enzyme Mechanism from Isotope Effects*; Cook, P.F., Ed.; CRC Press: Boca Raton, FL, 1991; pp 37-72.

[52] Hunkeler, D., Andersen, N., Aravena, R., Bernasconi, S. M., Butler, B. J. (2001). Hydrogen and carbon isotope fractionation during aerobic biodegradation of benzene. *Environ. Sci. Technol.*, 35, 3462–3467.

[53] Muneera, M., Theurichb, J., Bahnemannb, D. (2001). Titanium dioxide mediated photocatalytic degradation of 1,2-diethyl phthalate, *J. Photochem. Photobiol. A* , 143, 213–219.

[54] Balabanovich, A. I. and Schnabel, W. (1998). On the photolysis of phthalic acid dimethyl and diethylester: a product analysis study. J. Photochem. Photobiol., 113, 145–153.

[55] Lau, T. K., Chu, W., Graham, N. (2005). The degradation of endocrine disruptor di-n-butyl phthalate by UV irradiation: A photolysis and product study. *Chemosphere*, 60, 1045–1053.

[56] Shiner, V., J., Jr., Wilgis, F. P. Isotopes in organic chemistry. In *Heavy atom isotope effects*; Buncel, E., Saunders, W.H.,Jr., Eds.; Elsevier: New York, 1992; Vol. 8.

[57] Huang, D-S., Wu, S-H., Want, Y-S., Yu, S., Chan, S. (2002). Determination of the carbon kinetic isotope effects on propane hydroxylation mediated by the methane monooxyganases from methylococcus capsulatus (bath) by using stable carbon isotopic analysis. *Chem. Bio. Chem.*, 3, 760–765.

[58] Mancini, S. A., Devine, C. E., Elsner, M., Nandi, M. E., Ulrich, A. C., Edwards, E. A., Lollar, B. S. (2008). Isotopic evidence suggests different initial reaction mechanisms for anaerobic benzene biodegradation. *Environ. Sci. Technol.,* 42, 8290–8296.

[59] Fischer, A., Gehre, M., Breitfeld, J., Richnow, H.-H., Vogt, C. (2009). Carbon and hydrogen isotope fractionation of benzene during biodegradation under sulfate-reducing conditions: a laboratory to field site approach. *Rapid Commun. Mass Spectrom.*, 23, 2439–2447.

[60] Rakoczy, J., Remy, B., Vogt, C., Richnow, H. H. (2011). A bench-scale constructed wetland as a model to characterize benzene biodegradation processes in freshwater wetlands. *Environ. Sci. Technol.,* 45, 10036–10044.

[61] Fischer, A., Herklotz, I., Herrmann, S., Thullner, M., Weelink, S. A. B., Stmas, A. J. M., Schlomann, M., Richnow, H. H., Vogt, C. (2008). Combined carbon and hydrogen isotope fractionation investigations for elucidating benzene biodegradation pathways. *Environ. Sci. Technol.*, 42, 4356–4363.

[62] Nowak, J., Kirsch, N. H., Hegemann, W., Stan, H.-J. (1996). Total reductive dechlorination of chlorobenzenes to benzene by a methanogenic mixed culture enriched from the Saale river sediment. *Appl. Microbiol. Biotechnol.*, 45, 700–709.

In: Microcosms
Editor: Christopher C. Harris

ISBN: 978-1-62618-661-3

Chapter 5

BACTERIAL EVOLUTION IN SIMPLE MICROCOSMS

Andrew J. Spiers*

The SIMBIOS Centre and School of Contemporary Sciences,
University of Abertay Dundee, Dundee, UK

ABSTRACT

Microcosms have been used extensively to investigate ecological factors, evolutionary dynamics and the underlying molecular biology of adapting populations of bacteria in experimental evolution studies. One such experimental system has combined simple glass vials containing King's B liquid growth medium with the model soil and plant-associated pseudomonad, *Pseudomonas fluorescens* SBW25. In these, evolution reproducibly generates the Wrinkly Spreader genotype, a novel class of adaptive mutations that enable *Pf.* SBW25 to produce a robust, physical cohesive–class biofilm at the air-liquid interface. The analysis of Wrinkly Spreader mini-Tn*5* mutants led to the identification of cellulose as the main matrix component of the biofilm, and ultimately to the discovery of the mutation in a chemosensory-like system that activated the Wrinkly Spreader phenotype through the over-production of cyclic-*di*-GMP. Parallel ecological investigations have shown the importance of O_2 availability for the evolution and success of the Wrinkly Spreader. The initial colonisers establish an O_2 gradient, dividing the microcosm into a 100 – 200 μm high-O_2 zone at the air-liquid interface, and second, deeper low-O_2 zone in the rest of the liquid column. Growth is O_2–limited in King's B medium, and it is the domination of the air-liquid interface that guarantees access to O_2 and underlies the fitness advantage the Wrinkly Spreader has over non-biofilm–forming competitors including the ancestral wild-type *Pf.* SBW25. However, recent experiments demonstrate that Wrinkly Spreader fitness is reduced in static microcosms containing lower levels of nutrients, or in those in which King's B liquid viscosity has been increased. These findings demonstrate that the Wrinkly Spreader biofilm is a finely-balanced solution to the colonisation of the air-liquid interface in static microcosms, and changing environmental conditions will impact directly on Wrinkly Spreader fitness. This work highlights the importance of environmental factors, which along with competition, help guide the evolution of *Pf.* SBW25 in simple microcosms, invariably leading to the rise of the Wrinkly Spreaders.

* E-mail address: a.spiers@abertay.ac.uk.

INTRODUCTION

S.N. Winogradsky (1856 – 1953), founder of contemporary ecological microbiology, was one of the first to use simplified experimental environments or microcosms to investigate the physiology and behaviour of microorganisms [1, 16] (see also Chapter 1 of this book). In 1887 Winogradsky used a very simple microcosm formed under a glass cover slip with opposing O_2 and H_2S (hydrogen sulfide) gradients where he was able to demonstrate that the sulfur-oxidizing bacterium *Beggiatoa* had a preferred optimal zone for growth [80] (cited in [16]). More than 125 years later, microcosms are still being used to investigate microbiology. In experimental evolution studies, they are used to investigate ecological factors, dynamics and the underlying molecular biology of diversifying and adapting populations of bacteria (for reviews, see [13, 14, 15, 17, 36, 39, 40, 54, 72]). This is possible because random mutation, fast growth rates, short generation times, and high population densities provide ample variation for selection of adaptive genotypes (mutants with improved fitness). Furthermore, the ability to store ancestral and evolved isolates (strains or lineages) indefinitely allows comparative molecular analyses that can identify the mechanistic links between mutation and phenotype, and adaptation and environment. Although the microcosms used in experimental bacterial evolution studies may be simple, they provide ecological opportunity to drive diversification and adaptation (e.g. physical structure and chemical gradients defining different habitats, nutrients, temporal variation, etc.). Experimental conditions can also be manipulated to vary the relative importance of competition and environmental factors to the evolving populations (e.g. competition might be more important in high *cf.* low nutrient conditions). Here I describe the rise of the Wrinkly Spreader, an exemplar of experimental bacterial evolution studies. In this, the evolutionary dynamics occurring in diversifying populations of bacteria, the ecological basis for the adaptive advantage of the Wrinkly Spreader, and the underlying molecular biology, have all been investigated.

EXPERIMENTAL EVOLUTION WITH *P. FLUORESCENS* SBW25

The Wrinkly Spreader (WS) is both a genetically-defined mutant of the soil and plant-associated bacterium, *Pseudomonas fluorescens* SBW25, as well as the name given to a class of mutants exhibiting the same phenotype in experimental microcosms and on agar plates. *P. fluorescens* SBW25 (henceforth referred to as *Pf.* SBW25) was originally isolated from the leaf of a sugar beet (*Beta vulgaris*) [53]; the genome is fully sequenced [63] and the strain is genetically malleable using a wide range of techniques established for related pseudomonads, e.g. *P. aeruginosa, P. putida* and *P. syringae*. It is now regarded as a model organism for both environmental microbiology and experimental evolution. The rapid, reproducible diversification and adaptation of *Pf.* SBW25 populations in simple microcosms, as first described by Paul Rainey and Michael Travisano [55], has provided a system for experimental evolution studies. The evolutionary dynamics seen in these microcosms is explained by the Red Queen hypothesis which proposes that organisms must constantly evolve to gain reproductive advantage, but also to survive against ever-evolving competitors in an ever-changing environment [76] (however, see the Black Queen alternative [51, 59]).

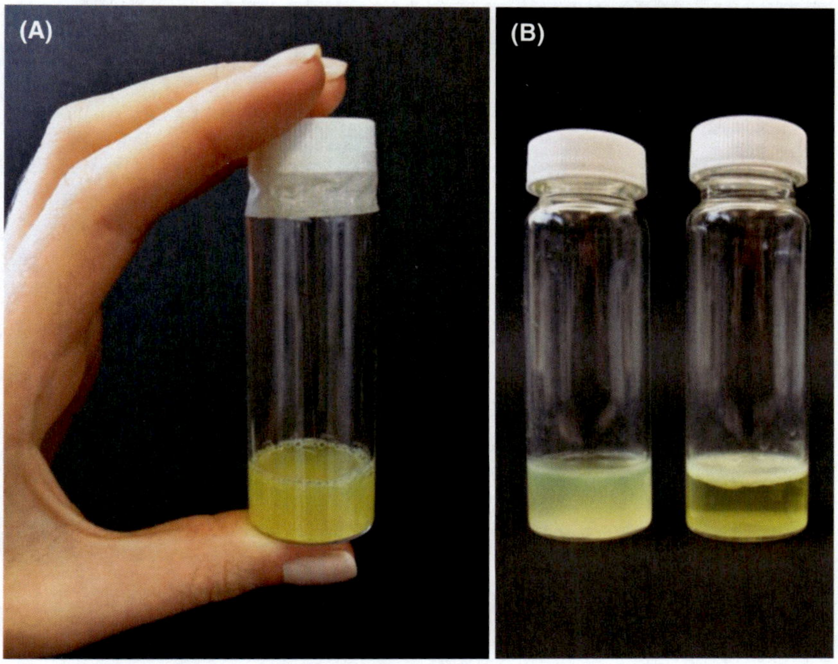

Figure 1. Experimental evolution in microcosms. The evolution of *Pf.* SBW25 has been investigated using simple microcosms consisting of (A) small glass vials containing King's B medium that are incubated statically or with shaking (the 6 ml liquid column has a surface area of 3.8 cm^2, a diameter of 2.2 cm, and a depth of 1.6 cm). (B) Wild-type or ancestral *Pf.* SBW25 (left) grows throughout the liquid column of static microcosms, whilst the Wrinkly Spreader (right) produces a biofilm at the air-liquid (A-L) interface and shows little growth through the liquid column. Source: Adapted from Green et al. [24].

The constant diversification and adaptation of *Pf.* SBW25 has allowed experimental evolution studies to investigate adaptive radiation in heterogeneous environments [55]; resource competition, nutrients and physical disturbance [10, 11, 43, 73], diversification [6, 12, 49, 50]; the evolution of co-operation [7, 8], community assembly [33], co-evolution of bacteria with phage and antibiotics [9, 19], pleiotropic fitness trade-offs [42], fitness distributions [46], and convergent phenotypic evolution [41] (this is not an exhaustive list of all publications; for reviews, see [13, 14, 15, 17, 36, 39, 40, 54, 72]). The simple microcosm system [55] consists of a 30 ml loosely-lidded glass vial containing 6 ml of nutritious King's B liquid growth medium, shown in Figure 1 (20 g Proteose peptone No. 3, 10 g glycerol, 1.5 g K_2HP0_4 and 1.5g $MgSO_4$ per litre [32]).

These are incubated with shaking to provide a homogenous environment, or statically without physical disturbance to provide a heterogeneous environment (i.e. no air-liquid interface or bottom surface, and no nutrient or O_2 gradients). In static microcosms, wild-type *Pf.* SBW25 colonists rapidly establish an O_2 gradient which defines three niches for colonisation and adaptation: the air-liquid (A-L) interface, the liquid column, and the vial bottom (microcosms subject to constant mixing do not develop different niches) [35]. Over a period of several days, typically within ~100 generations, wild-type *Pf.* SBW25 radiates or diversifies to produce a range of phenotypically distinguishable mutants to occupy all three niches [55].

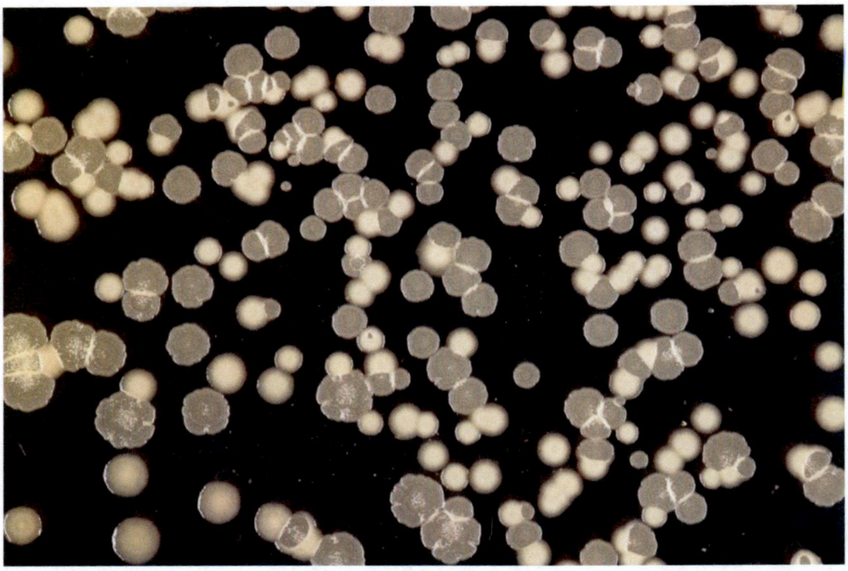

Figure 2. Diversification of wild-type *Pf.* SBW25 in static microcosms leads to a range of novel morphotypes. Shown is a sample from a static microcosm spread onto a King's B plate and incubated for several days. Wild-type or ancestral *Pf.* SBW25 produces round, smooth-looking colonies, and the Wrinkly Spreader produces flatter, wrinkled colonies. Source: Adapted from Green et al. [24].

These mutants are also referred to as morphotypes to reflect the variation seen in colony morphology on agar plates (sometimes abbreviated as morphs and used synonymously with mutant), and includes the Smooth, Wrinkly Spreader, and Fuzzy Spreader morphotypes. The Smooth morphotypes include the ancestral wild-type *Pf.* SBW25, and they produce smooth, rounded colonies and colonise the liquid column of static microcosms. The WS morphotypes produce wrinkled colonies (see Figure 2) and colonise the A-L interface through the formation of a robust biofilm. The Fuzzy Spreader morphotypes produce stippled-topped colonies and are thought to colonise the anoxic bottom of static microcosms [55] (an alternate explanation, as yet unpublished, is that they form a very fragile biofilm at the A-L interface that invariably fails, and as a result, Fuzzy Spreaders are always recovered from the bottom of the microcosm[1]).

RISE OF THE WRINKLY SPREADER

In experimental evolution studies, static microcosms are typically inoculated with $\sim 10^4$ wild-type *Pf.* SBW25 cells/ml and incubated for 5 days. Over this period, the population grows to $\sim 10^{10}$ cells/ml, during which the proportion of Wrinkly Spreader morphotypes increases from 0% to ~30% (e.g. [24]; see Figure 3) (WS-like mutants also occur in shaken microcosms but they never dominate the growing population and cannot form a biofilm). However, the successful domination of static microcosms by the Wrinkly Spreader is not due to a significantly faster rate of growth [35]. From the third day onwards it is quite common to

[1] Anna Koza, personal communication to Andrew Spiers, referring to a presentation given by Paul Rainey at the Department of Systems Biology of the Technical University of Denmark, 2012.

observe the formation of an A-L interface biofilm, presumably largely the result of the activity of the growing community of Wrinkly Spreader morphotypes (different WS mutants are known to arise in the same microcosm and can be differentiated on the basis of colony size and degree of wrinkleality or wrinkledness). It is probably the development of the biofilm which enables the Wrinkly Spreaders to compete more successfully than the non-biofilm–forming competitors for O_2 diffusing through the A-L interface into the liquid column below. Competitive fitness experiments have demonstrated that the archetypal Wrinkly Spreader (i.e., a specific strain recorded as PR1200 in [69] and LSWS in [2]) has a significant fitness advantage over non-biofilm–forming strains including wild-type *Pf.* SBW25 in static microcosms, but not in shaken microcosms or on agar plates where the WS phenotype is unstable [24, 55, 65, 69].

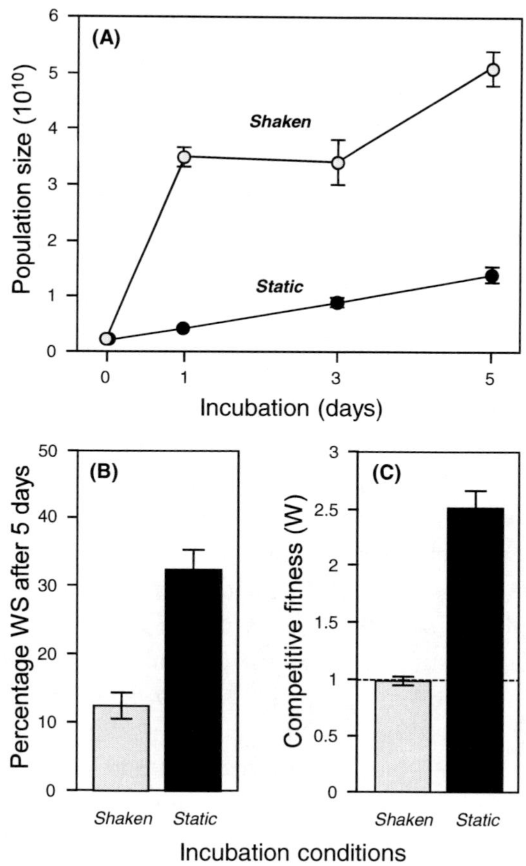

Figure 3. Rise of the Wrinkly Spreader in static microcosms. Morphotypes such as the Wrinkly Spreader accumulate in an expanding *Pf.* SBW25 populations in static microcosms over time. (A) Population growth is faster in shaken liquid microcosms than in static microcosms due to better aeration. (B) After five days, the percentage of Wrinkly Spreaders is higher in static microcosms than in shaken microcosms. (C) The Wrinkly Spreader has a competitive fitness (W) advantage compared to the non-biofilm–forming ancestral *Pf.* SBW25 in static microcosms where biofilms can occur but not in shaken microcosms where there is no advantage to biofilm formation. The dashed line indicates when $W = 1$ and the WS has no fitness advantage over the competitor; for $W > 1$ the WS has an advantage and for $W < 1$ the competitor has the advantage. Means and SE are shown. Source: Adapted from Green et al. [24].

Table 1. Radiation of *Pf.* SBW25 in static King's B microcosms

Strain / mutant	Niche preference	A-L Interface biofilm	Fitness compared to the ancestor in:	
			Static microcosms	Shaken microcosms
Ancestral wild-type Pf. SBW25	Liquid column	No		
Fuzzy Spreader	Vial bottom	No	High	Low
Smooth	Liquid column	No	Similar	Similar
Wrinkly Spreader	A-L interface	Yes	High	Low

A-L, air-liquid; Pf., Pseudomonas fluorescens.

This fitness advantage over the ancestral strain proves that the Wrinkly Spreader is not merely a novel mutant in terms of the molecular biology of *Pf.* SBW25, but that it is of evolutionary significance in which *Pf.* SBW25 has undergone an adaptive leap to a novel evolved form capable of colonizing a new niche (Table 1). In order to provide a mechanistic explanation of the WS phenotype, the underlying molecular biology was investigated. The Wrinkly Spreader colony morphology is similar to the rdar (rough, dry and rugose) colony morphology described for *Escherichia* and *Salmonella* spp. (reviewed by [58]), and like these, the Wrinkly Spreader produces a visually-obvious and robust biofilm at the air-liquid interface of static King's B microcosms. The ability to produce A-L interface biofilms in static microcosms is common amongst environmental pseudomonads, including water, soil and plant-associated, plant and mushroom pathogens, and psychrotrophic meat-spoiling pseudomonads [57, 66, 74] (see also the review of cellulose expression in *Pf.* SBW25 and other environmental pseudomonads by Spiers et al. [68]), and amongst pathogenic, environmental and produce-related *Enterobacteriaceae* [64, 81]. The formation of biofilms by bacteria is a key strategy in the colonisation of natural environments, though they are only one of a range of assemblages which range from isolated surface–attached bacteria, micro-colonies, and multi-layered, multi-species and differentiated biofilms, as well as flocs and slime (for recent biofilm reviews see [3, 18, 25, 48, 56, 78]).

Naturally-occurring multi-species biofilms are probably formed by a community of species that each contribute differently to the biofilm structure, nutrient acquisition and transformation, and the expression of anti-predatory and anti-competitor chemicals. In contrast to the generally-held view of biofilm formation as a genetically-determined developmental programme [52], biofilms are transient and *ad hoc* assembled communities better described by adaptation, social evolution and ecological succession [3, 36, 79]. Although the Red Queen may play a role driving competition and adaptation in biofilms, the Black Queen may play a greater role in developing longer-lasting and more robust interdependent co-operative communities [51, 59].

AIR-LIQUID INTERFACE BIOFILMS

The A-L interface biofilms produced by environmental pseudomonads have been categorised into the floccular mass, physically cohesive, viscous mass, and waxy aggregate biofilm classes on the basis of visual phenotype and the resistance to physical disturbance

[74], whilst the quantitative measurements of growth, biofilm strength and levels of attachment to the microcosm walls have allowed similar biofilms to be categorised into air-liquid, air-liquid-solid and liquid-solid interface biofilm types [57] (see Figure 4). Biofilms of the air-liquid interface type appear to 'float' on the liquid surface and have no significant attachment at the meniscus.

Like the air-liquid interface biofilm type, biofilms of the air-liquid-solid interface type colonise the liquid surface, but they also show significant levels of attachment to the microcosm vial walls at the meniscus. The air-liquid-solid biofilm type includes the physically cohesive–class biofilms, and are typified by the Wrinkly Spreader biofilm. In contrast to both the air-liquid and air-liquid-solid biofilm types, the liquid-solid interface biofilm type does not extend out across the liquid surface, and only colonises the meniscus region through attachment to the microcosm vial walls. This type does not include any of the biofilm categories previously described [74]. An analysis of biofilms produced by psychrotrophic pseudomonads recovered from spoilt meat suggests that the biofilm classes and types are in broad agreement [57], and that the value of biofilm typing is that it can quantify biofilms that are too insubstantial to be seen by eye (i.e. invisible biofilms resulting from very slow growth, no production of pigments or lacking contrast with the underlying culture growing in the liquid column).

A MECHANISTIC EXPLANATION FOR THE WS PHENOTYPE

The distinctive wrinkled colony morphology of the *Pf.* SBW25 Wrinkly Spreader mutant on King's B agar plates allowed the identification of the major loci involved in the WS phenotype by mini-transposon mutagenesis [2, 67, 69, 70].

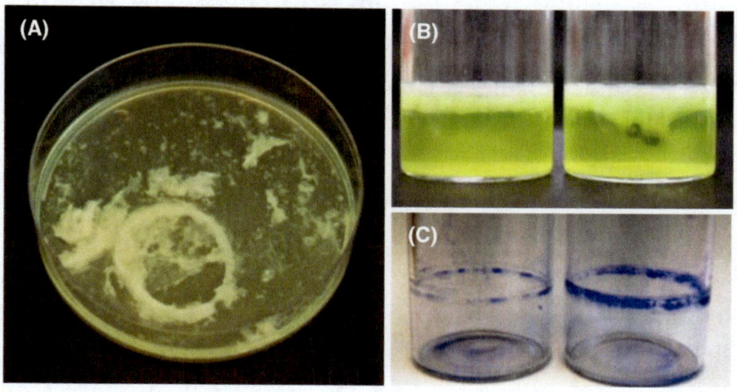

Figure 4. Measuring the robust Wrinkly Spreader biofilm. The success of the Wrinkly Spreader biofilm is in part due to its strength and the strong attachment to the glass walls of the microcosm. (A) The robustness of the Wrinkly Spreader biofilm is evident when the contents of the microcosm are poured into a petri dish where it remains largely intact. (B) Biofilm strength can be quantitatively measured using small glass placed onto the biofilm surface using forceps until the biofilm breaks or sinks. (C) Crystal violet-staining can be used to visualise attachment to the glass walls at the meniscus: wild-type *Pf.* SBW25 (left) shows relatively little attachment compared to the Wrinkly Spreader (right) (the stain can be eluted from the microcosms with ethanol and measured by spectrophotometry to provide a quantitative measure of attachment). Source: Adapted from Green et al. [24].

In this approach, a library of Wrinkly Spreader mutants containing the mini-Tn*5* transposon were screened for colonies showing a wild-type or Smooth–like morphology and were subsequently tested to see whether they still produced WS biofilms in static microcosms. Most of the WS mini-Tn*5* mutants identified in this manner were grouped into the wrinkly spreader structural (*wss*) and wrinkly spreader phenotype (*wsp*) operons [69], and collectively provide the mechanistic explanation of the WS phenotype. Key to this investigation was the ability to test the phenotype of all strains and mutants in static microcosms identical to those used to isolate the original Wrinkly Spreader mutant. The *wss* operon consists of ten genes (*wssA-J*) and shows a strong similarity to the bacterial cellulose synthetic (*bcs*) clusters originally identified as the *acs* operon in *Acetobacter xylinus* (now known as *Gluconacetobacter hansenii* ATCC 23769) and the *yhj* operon in *Escherichia coli* K-12 [4, 60].

The degree of amino acid-sequence homology between the *wss*, *acs* and *yhj* genes strongly suggested that the *wss* operon encoded a cellulose synthase complex, with WssA and J responsible for the correct localisation of the complex at the cell poles (though WssJ may be functionally redundant), WssBCD and E forming the core synthase, and WssFGHI partially acetylating the cellulose polymer [21, 38, 67, 69, 70, 71]. The production of cellulose was confirmed by various means, including fluorescent microscopy and the structural analysis of purified WS biofilm matrix material [67, 69, 70]. Fluorescent and scanning electron microscopy showed that the WS biofilm was dominated by an extensive network of extracellular cellulose fibres forming large clumps of material and thin films, with bacterial cells associated with the fibres and found within the voids (see Figure 5) [24, 35, 67]. Although the WS biofilm is > 97% by weight liquid, cellulose gels are notable for their ability to retain significant amounts of water [29, 62]. The structural integrity of WS biofilms could be destroyed by digestion with cellulases, indicating that cellulose was the major matrix component of the biofilm, though lipopolysaccharides on the surface of cells or associated with cell debris further strengthens the biofilm structure [67, 69, 70]. In related work, the expression of cellulose by wild-type *Pf.* SBW25 has been shown to be of advantage in the phytosphere of sugar beet (*Beta vulgaris*) seedlings [20] (the *wss* operon was also identified by IVET (*in vivo* expression technology) analysis (reviewed by [30]) to be specifically induced by the sugar beet rhizosphere [20]), and on mushroom (*Agaricus bisporus*) caps where it may increase survival by reducing the impact of water stress[2].

The *wsp* operon consists of seven genes (*wspA-F and R*) [2] and shows strong similarity to the *E. coli* Che chemosensory system (reviewed by [5]). WspABD forms a membrane-bound chemoreceptor regulated by the antagonistic WspC methytransferase and WspF methylesterase. In the absence of an as-yet unidentified environmental signal, the chemoreceptor is silent and does not activate the associated response regulator, WspR, by phosphorylation. However, in the Wrinkly Spreader mutant (the genotype is wild-type *Pf.* SBW25 but *wspF* S301R [2]), WspF function is reduced or lost, leading to the inappropriate activation of WspR in the absence of environmental signal. WspR is a DGC (di-guanylate cyclase), and the phosphorylated active form synthesizes the intracellular signal cyclic-*di*-GMP (bis-(3'-5')-cyclic dimeric guanosine monophosphate) from GTP [23, 45] (for a recent review of cyclic-*di*-GMP signalling in bacteria, see [31, 61]).

[2] Anna Koza and Andrew Spiers, manuscript in preparation.

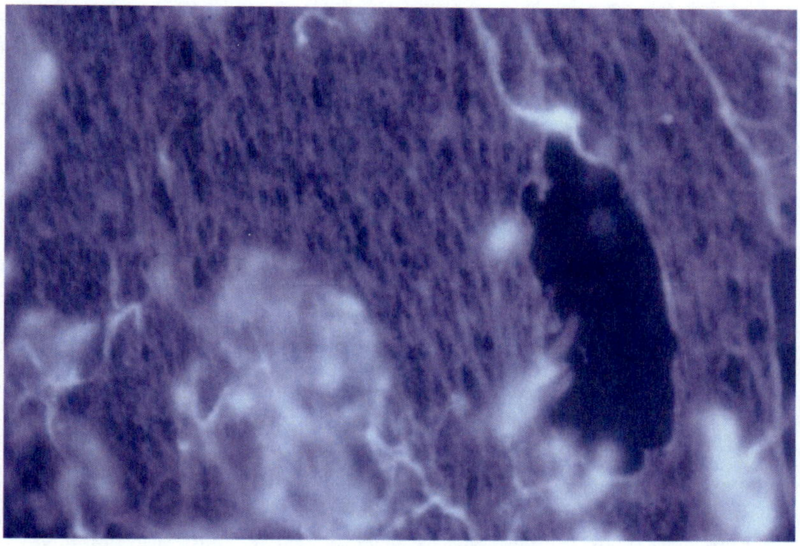

Figure 5. The Wrinkly Spreader biofilm matrix is a network of cellulose fibres, clumps and films. Shown is a low-magnification (10x) fluorescent image of a sample of the Wrinkly Spreader biofilm after staining with Calcofluor to image cellulose. Higher-levels of magnification are required to visualise individual bacterial cells found attached to fibres and occupying the voids between fibres. Source: Adapted from Green et al. [24].

Table 2. Attributes of the Wrinkly Spreader

Microcosm niche	Air-liquid interface.
Morphotype / Genotype	Wrinkly Spreader / Wild-type but *wspF* S301R.
Consequence	WspF subunit function reduced or destroyed, resulting in the activation of the DGC WspR and increased levels of *c-di*-GMP. Increased levels of *c-di*-GMP activates the cellulose synthase and also induces the expression of an unidentified attachment factor, producing a biofilm.
	Niche changed from liquid column to the air-liquid interface.
	Increased fitness in static microcosms relative to non-biofilm-forming competitors including the ancestral wild-type *Pf.* SBW25.
Biofilm	Robust, substantial, and well-attached at the meniscus to the microcosm vial walls.
Biofilm class	Physically cohesive (PC).
Biofilm type	Air-liquid-solid interface (A-L-S).
Matrix	Partially-acetylated cellulose, attachment factor and lipopolysaccharide.

DGC, di-guanylate cyclase; Pf., Pseudomonas fluorescens.

Increases in the level of c-*di*-GMP leads to the activation of the Wrinkly Spreader phenotype through the direct stimulation of the cellulose synthase complex and provides a mechanistic explanation of the Wrinkly Spreader phenotype (Table 2; see Figure 6) [2] (the second component required for the WS phenotype, the pilli-like attachment factor, has not yet been identified and it is not known how it might be regulated by c-*di*-GMP).

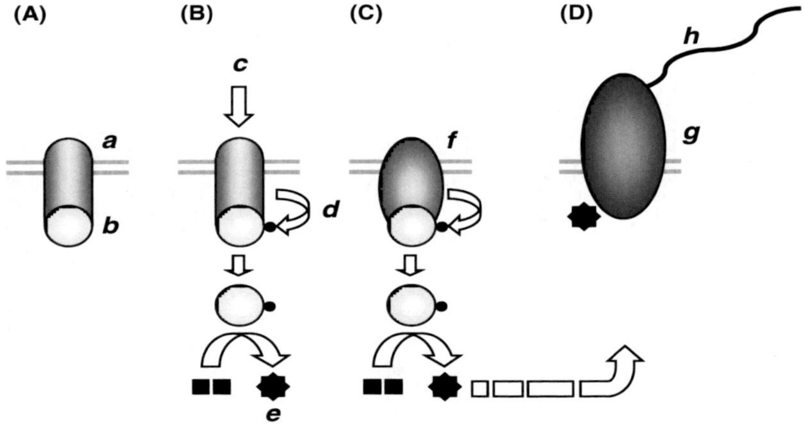

Figure 6. A mechanistic explanation for the Wrinkly Spreader phenotype. The activation pathway in the Wrinkly Spreader is shown schematically. (A) Under normal circumstances the Wsp chemosensory complex (a) and associated WspR regulator (b) are inactive. (B) When the appropriate environmental signal (c) is received, the Wsp complex activates WspR by phosphorylation (d) which produces the secondary messenger c-*di*-GMP (e) from GMP. (C) In the Wrinkly Spreader, a mutation in the Wsp complex (f) leads to the constant activation of WspR and production of c-*di*-GMP. This activates cellulose synthase (g) to produce cellulose (h), a central element of the Wrinkly Spreader phenotype which includes the biofilm and wrinkled colony morphology. Source: Adapted from Bantinaki et al. [2] and Green et al. [24].

A number of mutations in *wspF* have now been identified that activate the WS phenotype, as well as mutations associated with other DGCs which are also thought to increase c-*di*-GMP levels [2, 21, 47]. Finally, in an extension of the IVET analysis, AlgR, a regulator of alginate expression, and FleQ, a regulator of flagella expression, were also identified as regulators of the *wss* operon [22] in addition to WspR previously identified as being involved [69], suggesting that cellulose expression may also be regulated at the transcriptional level in response to high-level cellular regulators and environmental signals. In the WS, *wss* transcriptional levels are higher than in the wild-type *Pf.* SBW25, but an increase in *wss* transcription is not sufficient to explain the WS phenotype [69] as this does not also result in the increase of c-*di*-GMP levels.

Establishing the Wrinkly Spreader as an Adaptive Genotype

Competitive fitness assays were used to establish the improved fitness of the Wrinkly Spreader compared to non-biofilm–forming strains and the ancestral *Pf.* SBW25. Competitive fitness (W) was estimated from the ratio of the Malthusian parameters (m) for the Wrinkly Spreader and the non-biofilm–forming reference (i.e. $m_{WS} / m_{Reference}$), where $m = ln\ [N_f / N_0]$ and N_0 and N_f are the initial and final numbers of each strain over the period of the assay [37]. When $W > 1$, the Wrinkly Spreader has a fitness advantage over the reference strain; $W = 1$, the two strains are neutral, and when $W < 1$, the reference strain has the advantage over the Wrinkly Spreader. In static microcosms in which the Wrinkly Spreader can form a biofilm, it has a significant fitness advantage ($W = 1.7 - 2.5$; see Figure 3(b) [24, 69]). In contrast, in shaken microcosms where no biofilm can form, the Wrinkly Spreader fitness is poor compared to the reference strains ($W = 0.33 - 1.00$), and on agar plates on which the Wrinkly Spreader is genetically unstable and produces Smooth-like revertants within days, the fitness is even lower ($W = 0.13 - 0.83$) [24, 65, 69]. The evolutionary and ecological explanation of the fitness advantage of the Wrinkly Spreader in static microcosms is that the benefit of A-L interface colonisation outweighs the cost of biofilm construction and maintenance. The benefit is a faster growth due to greater O_2 availability at the A-L interface compared to that achieved in deeper regions of the microcosm [35] (the role of O_2 is discussed in the next section), whilst the cost is the energy diverted from growth required to produce a structure sufficiently robust to overcome minor physical disturbance occurring during incubation. This disturbance appears to be the main environmental constraint on biofilm formation, as *Pf.* SBW25 growth is not limited by nutrient availability in the King's B static microcosms [67].

Colonisers Generate Ecological Opportunities

The mechanistic explanation of the rise of the Wrinkly Spreader in static microcosms describes how a single mutation in a regulatory subunit of a chemosensory complex can result in the production of a robust biofilm at the air-liquid interface of static microcosms, whilst fitness assays demonstrate that such biofilm–forming mutants have a higher Darwinian fitness than their non-biofilm–forming competitors and the ancestral *Pf.* SBW25. However, the mechanistic explanation and the improved fitness are not sufficient to explain the rise of the Wrinkly Spreaders *per se*: an ecological perspective is required to provide the selective pressure that favours the Wrinkly Spreader over all other morphotypes in the static microcosm. Although the *wspF* S301R mutation is the proximate cause of the WS phenotype, the success of the Wrinkly Spreader was attributed to the greater availability of O_2 at the air-liquid interface where it produces a robust biofilm [55]. However, the importance of O_2 gradients in *Pf.* SBW25 adaptation was not examined until the explicit testing of the role of O_2 in evolving populations using micro-sensor profiling and experiments conducted under high and low–O_2 conditions [35].

In this, the initial wild-type *Pf.* SBW25 colonists of static microcosms were found to establish O_2 gradients over 2 – 3 hours before significant population growth had occurred, converting a previously homogenous environment into one containing a resource continuum with high and low–O_2 regions [35] (see Figures 7 and 8). The high–O_2 region corresponds to a 100 – 200 μm layer immediately below the air-liquid interface, whilst the low–O_2 region occupies the rest of the 1.6 cm deep liquid column in the static microcosm. *Pf.* SBW25 growth in King's B medium is O_2-limited, and in static microcosms, growth is probably the greatest in the high-O_2 region near the A-L interface [35]. Furthermore, the O_2 gradients established by the early colonists persisted over several days during which time significant numbers of Wrinkly Spreaders appear by random mutation.

Different O_2 conditions have a significant impact on the metabolism and physiology of aerobic and facultative anaerobic bacteria (most *P. fluorescens* strains are regarded as obligate aerobes: *Pf.* SBW25 will grow under micro-aerobic conditions in King's B with only 0.5% of normal levels of O_2 [35]), and this has been investigated by single-cell Raman spectroscopy of WS cells [28]. Raman analysis can be used to provide a spectral fingerprint of individual cells and is generally used to differentiate between bacterial strains [44]. However, it can also demonstrate differences in metabolism and physiology for a specific bacterium growing on different defined media or in different environmental conditions (e.g. *Pf.* SBW25 growing on glucose or fructose, arginine or aspartic acid, King's B, plant-derived media, on leaves or under starvation conditions, with a pseudomonas-associated environmental plasmid, etc.) [26, 27, 75]. Raman spectral analysis could differentiate between WS cells growing under high and low–O_2 conditions, suggesting that O_2 levels had a significant impact on metabolism and physiology [28]. It also differentiated between cells recovered from the biofilm and deeper down the liquid column, again suggesting that metabolic and physiological shifts had occurred in response to changing environmental conditions [28]. Under high–O_2 conditions, the fitness of the Wrinkly Spreader compared to a non-biofilm–forming reference strain was high (W = 1.2), but under low–O_2 conditions, the fitness was significantly reduced (W = 0.1), indicating that access and uptake of O_2 was the ecological prize achieved by forming a biofilm at the air-liquid interface of static microcosms [35].

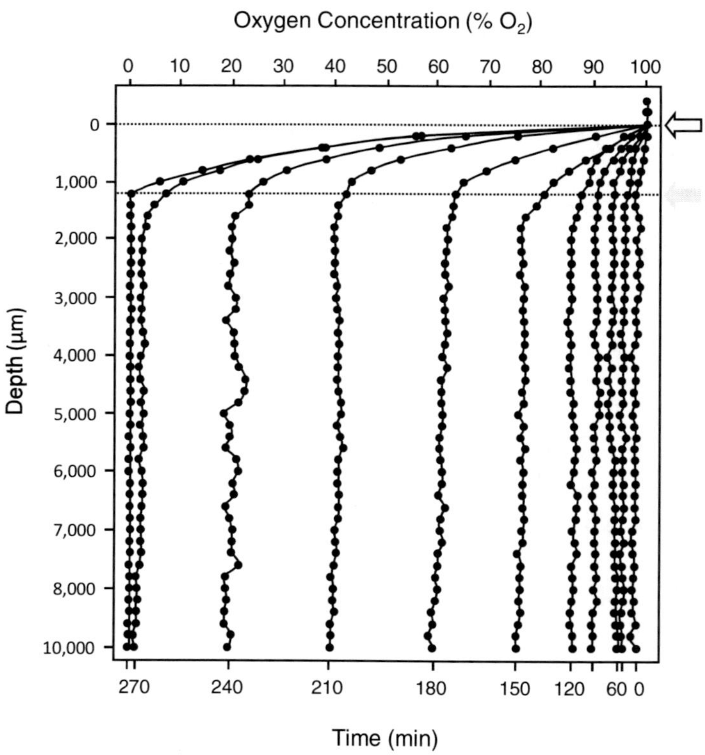

Figure 7. *Pf.* SBW25 colonists in static microcosms rapidly establish an O_2 gradient. Micro-probe profiling was used to measure the development of O_2 gradients in static microcosms inoculated with *Pf.* SBW25. Measurements were made every 200 µm down a profile of 10 mm, and profiles 30 min apart are shown. The white arrow indicates the air-liquid interface; microcosms were ~16 mm deep and the final 6 mm were not profiled. O_2 concentration is indicated as the percentage of normal O_2 levels in solution (% O_2) on the top x-axis. Total elapsed time since inoculation is indicated on the bottom x-axis. The grey arrow indicates ≤ 1% O_2 reached after 300 min at a depth of 1.2 mm. Source: Figure from Koza et al. [35].

CHALLENGING THE FITNESS OF THE WRINKLY SPREADER

In static microcosms, O_2 availability is the dominant environmental factor driving the evolution and fitness of the Wrinkly Spreader. It accesses O_2 through the formation of a robust biofilm at the A-L interface, and this adaptive solution clearly provides a significant benefit over costs as the fitness of the Wrinkly Spreader is always greater than that of competing non-biofilm–forming strains. However, the conditions in static microcosms can be altered in such a manner as to reduce the fitness advantage of the Wrinkly Spreader, illustrating how the interaction of different environmental factors or initial physical conditions can influence the outcome of bacterial evolution adaptation. Recent experiments[3] suggest that as the relative level of nutrients present in standard King's B microcosms is reduced, the competitive fitness of the Wrinkly Spreader falls (Figure 9). Although growth rates are lower in reduced-nutrient media, the Wrinkly Spreader is still able to produce a biofilm in 0.01x King's B nutrients static microcosms. However, the costs of producing a biofilm under these conditions are considerably higher than the advantage gained by biofilm–formation. The evolution of Wrinkly Spreaders appears less certain when the relative importance of O_2 and nutrients are altered in static microcosms. Analysis of nutrient and O_2 gradients in air-liquid biofilms produced by *Gluconacetobacter xylinus* (formerly known as *Acetobacter xylinum*) suggest that both limit growth, with O_2 levels falling below optimal levels at depths greater than 100 μm, and nutrient levels falling below optimal levels at depths less than 50 μm, to define an optimal growth zone 50 – 100 μm below the air-liquid interface [77]. Although nutrient gradients have not been measured in evolving *Pf.* SBW25 static microcosms or in Wrinkly Spreader biofilms, we would expect to see a similar interaction between O_2 and nutrient gradients defining zones of optimal and sub-optimal growth which in turn provide ecological opportunity for colonisation and further adaptation.

A second mutant known as the CBFS (Complementary Biofilm Forming Strain) is also able to form a biofilm in static microcosms. The CBFS was evolved from a *Pf.* SBW25 cellulose-deficient strain [21] and produces a robust, physically cohesive–class biofilm. In addition, when induced by exogenous iron, wild-type *Pf.* SBW25 produces the VM (Viscous Mass) biofilm in static microcosms. In this fragile viscous mass–class biofilm, cellulose is the major matrix component [34]. A quantitative comparison of the CBFS, VM and WS biofilms indicates that the three biofilms are physically distinct structures, and all three provide similar benefits over a common non-biofilm–forming strain as determined by competitive fitness assays[4]. The function of each biofilm is to retain the resident bacteria at the air-liquid interface, and preferably in the optimal high–O_2 and nutrient zone in which maximal growth can be achieved. The physical structure of the biofilms might reduce the random diffusion of bacterial cells by trapping them in a fine-mesh cellulose–matrix network or by allowing temporary attachment to larger matrix fibres. Within the optimal zone individual cells could manoeuvre by aero- and chemotaxis; *Pf.* SBW25 swimming (flagella-mediated) motility has been characterised in soft-agar plate assays and in liquids [70]. Presumably the cost of constantly maintaining position by swimming outweighs the cost of producing a biofilm to restrict cell diffusion, when assessed at the population level and over the time needed for biofilm–formation.

[3] Ella Clapperton and Andrew Spiers; to be published elsewhere in full at a later date.
[4] Anna Koza and Andrew Spiers, manuscript in preparation.

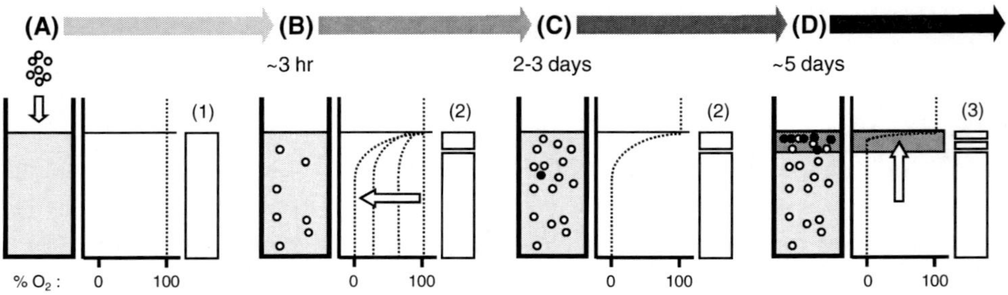

Figure 8. The colonists of static microcosms modify the original environment and generate new niches for their descendants. Shown are static microcosms with growing and diversifying *Pf.* SBW25 populations (circles), corresponding O_2 gradients (dashed line) and niches (rectangular boxes). (A) Wild-type *Pf.* SBW25 colonists are added to a homogenous environment with no O_2 gradient (i.e. a single niche). (B) Within 3 h the colonists have established a significant O_2 gradient, modifying the original environment to produce a resource continuum containing a high–O_2 zone at the A-L interface and a low–O_2 region lower down the liquid column. (C) Over the next 2-3 days, significant population expansion and diversification occur, during which Wrinkly Spreader-like morphotypes appear (black circle). (D) In the following days, Wrinkly Spreaders begin to dominate the A-L interface by producing a biofilm (grey rectangle). As the biofilm grows in depth, the transition zone between high and low–O_2 environments will become fixed near the A-L interface, establishing physically-structured high and low–O_2 niches within the biofilm (a third unstructured low–O_2 niche is below the biofilm), in which further adaptation and succession can occur. Source: Figure from Koza et al. [35].

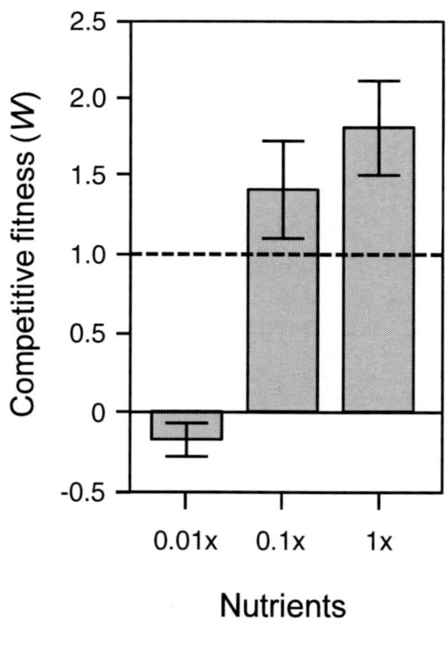

Figure 9. Wrinkly Spreader fitness in static microcosms is reduced as nutrient levels are lowered. Shown are preliminary results of experiments determining the competitive fitness (*W*) of the Wrinkly Spreader compared to a non–biofilm–forming reference strain in static King's B microcosms containing 0.01x, 0.1x and 1x normal levels of nutrients. Means and standard errors are shown. The dashed line indicates when $W = 1$ and the WS has no fitness advantage over the competitor; for $W > 1$ the WS has an advantage and for $W < 1$ the competitor has the advantage. Although calculations might result in $W < 1$, it is explicitly presumed that $W > 0$ [37]; in the case of the 0.01x nutrients here, the mean is not significantly different to zero (t-test, $t_{7,\,-1.6471}$, $p = 0.1435$). Source: Ella Clapperton and Andrew Spiers.

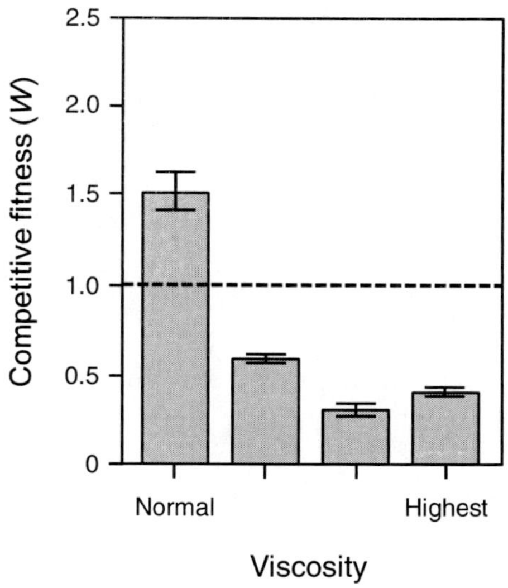

Figure 10. Wrinkly Spreader fitness in static microcosms is reduced as viscosity is increased. Shown are preliminary results of experiments determining the competitive fitness (*W*) of the Wrinkly Spreader compared to a non–biofilm–forming reference strain in static King's B microcosms in which liquid viscosity has been increased with the addition of PEG (polyethylene glycol). The x-axis scale is arbitrary and does not reflect real viscosity values, but shows from left to right normal viscosity microcosms with no added PEG to the highest viscosity microcosm with 10% (w/v) PEG. Means and standard errors are shown. The dashed line indicates when $W = 1$ and the WS has no fitness advantage over the competitor; for $W > 1$ the WS has an advantage and for $W < 1$ the competitor has the advantage. Source: Denise Barrett and Andrew Spiers.

This presumption is supported by preliminary fitness assays[5] in which the viscosity of the King's B medium has been increased to reduce cell diffusion. As viscosity is increased, the fitness of the Wrinkly Spreader is reduced compared to a non-biofilm–forming competitor strain (Figure 10). These findings suggest that there might be physical conditions where the costs of maintaining position by swimming might be lower than the cost of biofilm–formation. Under such conditions the Wrinkly Spreader might not dominate diversifying populations, and would be out-competed by efficient swimmer genotypes.

FINAL COMMENT

The *Pf.* SBW25 microcosm has been used extensively in experimental evolution to address questions in bacterial adaptation. The molecular analysis of one particular adaptive genotype reproducibly isolated from these microcosms, the Wrinkly Spreader, has led to a mechanistic explanation linking mutation with the WS phenotype, and recently, the role of O_2 gradients and availability within the microcosm has been shown to drive the evolution of the Wrinkly Spreader. However, recent investigations suggest that the relative levels of O_2 and nutrients, and the relative expense of cellulose production compared to swimming, are the real criteria underpinning the success of the Wrinkly Spreader. It is heartening to realise that, fifteen years after the first publication describing the diversification and adaptation of *Pf.*

[5] Denise Barrett and Andrew Spiers; to be published elsewhere in full at a later date.

SBW25, there is still more to be discovered by the combined evolutionary, ecological and molecular analysis of such a simple microcosm. These discoveries, like the work published to date, will not just be of interest to a small group of researchers, but will be of value to a wider audience of evolutionary ecologists and environmental microbiologists interested in understanding how organisms sense, respond to, and ultimately change their environment.

ACKNOWLEDGMENTS

I thank colleagues for their valuable comments on the manuscript. Andrew Spiers is a member of the Scottish Alliance for Geoscience Environment and Society (SAGES), and is supported by the SIMBIOS Centre and the University of Abertay Dundee. The University of Abertay Dundee is a charity registered in Scotland, No: SC016040.

REFERENCES

[1] Ackert, L.T. (2007). The "Cycle of Life" in ecology: Sergei Vinogradskii's soil microbiology, 1885–1940. *J. History Biol.* 40: 109–145.

[2] Bantinaki, B., Kassen, R., Knight, C.G., Robinson, Z., Spiers, A.J. and Rainey, P.B. (2007). Adaptive divergence in experimental populations of *Pseudomonas fluorescens.* III. Mutational origins of Wrinkly Spreader diversity. *Genetics* 176: 441–453.

[3] Battin, T., Sloan, W.T., Kjelleberg, S., Daims, H., Head, I.M., Curtis, T.P. and Eberl, L. (2007). Microbial landscapes: new paths to biofilm research. *Nature Rev. Microbiol.* 5: 76–81.

[4] Blattner, F.R., Plunkett, G., Bloch, C.A., Perna, N.T., Burland, V., Riley, M., Collado-Vides, J., Glasner, J.D., Rode, C.K., Mayhew, G.F., Gregor, J., Davis, N.W., Kirkpatrick, H.A., Goeden, M.A., Rose, D.J., Mau, B. and Shao, Y. (1997). The complete genome sequence of *Escherichia coli* K-12. *Science* 277: 1453-1462.

[5] Bren, A. and Eisenbach, M. (2000). How signals are heard during bacterial chemotaxis: protein-protein interactions in sensory signal propagation. *J. Bacteriol.* 182: 6865-6873.

[6] Brockhurst, M.A., Colegrave, N., Hodgson, D.J. and Buckling, A. (2007). Niche occupation limits adaptive radiation in experimental microcosms. *PLoS ONE* 2: e193.

[7] Brockhurst, M.A., Habets, M.G.J.L., Libberton, B., Buckling, A. and Gardner, A. (2010). Ecological drivers of the evolution of public-goods cooperation in bacteria. *Ecology* 91: 334–340.

[8] Brockhurst, M.A., Hochberg, M.E., Bell, T. and Buckling, A. (2006). Character displacement promotes cooperation in bacterial biofilms. *Curr. Biol.* 16: 1–5.

[9] Brockhurst, M.A., Morgan, A.D., Fenton, A. and Buckling, A. (2007). Experimental coevolution with bacteria and phage : The *Pseudomonas fluorescens*–ϕ2 model system. *Infect. Genet. Evol.* 7: 547–552.

[10] Buckling, A., Brockhurst, M.A., Travisano, M. and Rainey, P.B. (2007). Experimental adaptation to high and low quality environments under different scales of temporal variation. *J. Evol. Biol.* 20: 296–300.

[11] Buckling, A., Kassen, R., Bell, G. and Rainey, P.B. (2000). Disturbance and diversity in experimental microcosms. *Nature* 408: 961–964.

[12] Buckling, A., Wills, M.A. and Colegrave, N. (2003). Adaptation limits diversification of experimental bacterial populations. *Science* 302: 2107–2109.

[13] Buckling, A., Maclean, C.R., Brockhurst, M.A. and Colegrave, N. (2009). The Beagle in a bottle. *Nature* 457: 824–829.

[14] Cooper, T.F. (2012). Empirical insights into adaptive landscapes from bacterial experimental evolution. In: The Adaptive Landscape. Svensson, E.I. and Calsbeek, R. (Eds). Oxford University Press, Oxford.

[15] Dalziel, A.C., Rogers, S.M. and Schulte, P.M. (2009). Linking genotypes to phenotypes and fitness: how mechanistic biology can inform molecular ecology. *Mol. Ecol.* 18: 4997–5017.

[16] Dworkin, M. (2012). Sergei Winogradsky: a founder of modern microbiology and the first microbial ecologist. *FEMS Microbiol. Rev.* 36: 364–379.

[17] Elena, S.F. and Lenski, R.E. (2003). Evolution experiments with microorganisms: the dynamics and genetic bases of adaptation. *Nature Rev. Genet.* 4: 457–469.

[18] Elias, S. and Banin, E. (2012). Multi-species biofilms: living with friendly neighbors. *FEMS Microbiol. Rev.* 36: 990–1004.

[19] Escobar-Páramo, P., Gougat-Barbera, C. and Hochberg, M.E. (2012). Evolutionary dynamics of separate and combined exposure of *Pseudomonas fluorescens* SBW25 to antibiotics and bacteriophage. *Evol. Applications* 5: 583–592.

[20] Gal, M., Preston, G.M., Massey, R.C., Spiers, A.J. and Rainey, P.B. (2003). Genes encoding a cellulosic polymer contribute toward the ecological success of *Pseudomonas fluorescens* SBW25 on plant surfaces. *Mol. Ecol.* 12: 3109–3121.

[21] Gehrig, S.M. (2005). Adaptation of *Pseudomonas fluorescens* SBW25 to the air-liquid interface: a study in evolutionary genetics. DPhil thesis. University of Oxford, Oxford.

[22] Giddens, S.R., Jackson, R.W., Moon, C.D., Jacobs, M.A., Zhang, X-X., Gehrig, S.M. and Rainey, P.B. (2007). Mutational activation of niche-specific genes provides insight into regulatory networks and bacterial function in a complex environment. *Proc. Natl. Acad. Sci. (USA)* 104: 18247–18252.

[23] Goymer, P., Kahn, S.G., Malone, J.G., Gehrig, S.M., Spiers, A.J. and Rainey, P.B. (2006). Adaptive divergence in experimental populations of *Pseudomonas fluorescens*. II. Role of the GGDEF regulator WspR in evolution and development of the Wrinkly Spreader phenotype. *Genetics* 173: 515–526.

[24] Green, J.H., Koza, A., Moshynets, O., Pajor, R., Ritchie, M.R. and Spiers, A.J. (2011). Evolution in a test tube: rise of the Wrinkly Spreaders. *J. Biol. Educ.* 45: 54–59.

[25] Hall-Stoodley, L., Costerton, J.W. and Stoodley, P. (2004). Bacterial biofilms: from the natural environment to infectious diseases. *Nature Rev. Microbiol.* 2: 95–108.

[26] Huang, W.E., Bailey, M.J., Thompson, I.P., Whiteley, A.S. and Spiers, A.J. (2007). Single-cell Raman spectral profiles of *Pseudomonas fluorescens* SBW25 reflects *in vitro* and *in planta* metabolic history. *Microb. Ecol.* 53: 414–425.

[27] Huang, W.E. and Spiers, A.J. (2006). Consideration of future requirements for Raman microbiology as an examplar for the *ab initio* development of informatics frameworks for emergent OMIC technologies. *OMICS: J. Integ. Biol.* 10: 238–241.

[28] Huang, W.E., Ude, S. and Spiers, A.J. (2007). *Pseudomonas fluorescens* SBW25 biofilm and planktonic cells have differentiable Raman spectral profiles. *Microb. Ecol.* 53: 471–474.

[29] Hoshino, E., Wada, Y. and Nishizawa, K. (1999). Improvements in the hygroscopic properties of cotton cellulose by treatment with an endo-type cellulase from *Streptomyces* sp. KSM-26. *J. Biosci. Bioeng.* 88: 519–525.

[30] Jackson, R.W. and Giddens, S.R. (2006). Development and application of *in vivo* expression technology (IVET) for analysing microbial gene expression in complex environments. *Infect. Disord. Drug Targets* 6: 207–240.

[31] Jenal, U. and Malone, J. (2006). Mechanisms of cyclic-di-GMP signaling in bacteria. *Annu. Rev. Genet.* 40: 385–407.

[32] King, E.O., Ward, M.K. and Raney, D.C. (1954). Two simple media for the demonstration of pyocyanin and fluorescin. *J. Lab. Clin. Med.* 44: 301–307.

[33] Knope, M.L., Forde, S.E. and Fukami, T. (2012). Evolutionary history, immigration history, and the extent of diversification in community assembly. *Frontiers Microbiol.* 2: 273.

[34] Koza, A., Hallett, P.D., Moon, C.D. and Spiers, A.J. (2009). Characterization of a novel air–liquid interface biofilm of *Pseudomonas fluorescens* SBW25. *Microbiology* 155: 1397–1406.

[35] Koza, A., Moshynets, O., Otten, W. and Spiers, A.J. (2011). Environmental modification and niche construction: Developing O_2 gradients drive the evolution of the Wrinkly Spreader. *ISME J.* 5: 665–673.

[36] Kreft, J-U. and Bonhoeffer, S. (2005). The evolution of groups of cooperating bacteria and the growth rate versus yield trade-off. *Microbiology* 151: 637–641.

[37] Lenski, R.E., Rose, M.R., Simpson, S.C. and Tadler, S.C. (1991). Long-term experimental evolution in *Escherichia coli*. I. Adaptation and divergence during 2,000 generations. *Amer. Nat.* 138: 1315–1341.

[38] Le Quéré, B. and Ghigo, J.M. (2009). BcsQ is an essential component of the *Escherichia coli* cellulose biosynthesis apparatus that localizes at the bacterial cell pole. *Mol. Microbiol.* 72: 724–740.

[39] MacLean, R.C. (2005). Adaptive radiation in microbial microcosms. *J. Evol. Biol.* 18: 1376–1386.

[40] MacLean, R.C. (2008). The tragedy of the commons in microbial populations: insights from theoretical, comparative and experimental studies. *Heredity* 100: 471–477.

[41] MacLean, R.C. and Bell, G. (2003). Divergent evolution during an experimental adaptive radiation. *Proc. R. Soc. Lond. B* 270: 1645–1650.

[42] MacLean, R.C., Bell, G. and Rainey, P.B. (2004). The evolution of a pleiotropic fitness tradeoff in *Pseudomonas fluorescens*. *Proc. Natl. Acad. Sci. USA* 101: 8072–8077.

[43] MacLean, R.C., Dickson, A. and Bell, G. (2005). Resource competition and adaptive radiation in a microbial microcosm. *Ecol. Lett.* 8: 38–46.

[44] Magee, J. (1993). Whole-organism fingerprinting. In: Handbook of New Bacterial Systematics. Goodfellow, M. and O'Donnell, A.G. (Eds.). Harcourt Brace, New York.

[45] Malone, J.G., Williams, R., Christen, M., Jenal, U., Spiers, A.J. and Rainey, P.B. (2007). The structure–function relationship of WspR, a *Pseudomonas fluorescens* response regulator with a GGDEF output domain. *Microbiology* 153: 980–994.

[46] McDonald, M.J., Cooper, T.F., Beaumont, H.J.E. and Rainey, P.B. (2011). The distribution of fitness effects of new beneficial mutations in *Pseudomonas fluorescens. Biol. Lett.* 7: 98–100.

[47] McDonald, M.J., Gehrig, S.M., Meintjes, P.L., Zhang, X. and Rainey, P.B. (2009). Adaptive divergence in experimental populations of *Pseudomonas fluorescens*. IV. Genetic constraints guide evolutionary trajectories in a parallel adaptive radiation. *Genetics* 183: 1041-1053.

[48] McDougald, D., Rice, S.A., Barraud, N., Steinberg, P.D. and Kjelleberg, S. (2011). Should we stay or should we go: mechanisms and ecological consequences for biofilm dispersal. *Nature Rev. Microbiol.* 10: 39–50.

[49] Meyer, R. and Kassen, R. (2007). The effects of competition and predation on diversification in a model adaptive radiation. *Nature* 446: 432–435.

[50] Meyer, J.R., Schoustra, S.E., Lachapelle, J. and Kassen, R. (2011). Overshooting dynamics in a model adaptive radiation. *Proc. R. Soc. B* 278: 392-398.

[51] Morris, J.J., Lenski, R.E. and Zinser, E.R. (2012). The Black Queen Hypothesis: evolution of dependencies through adaptive gene loss. *mBio* 3: e00036-12.

[52] O'Toole, G., Kaplan, H.B. and Kolter, R. (2000). Biofilm formation as microbial development. *Annu. Rev. Microbiol.* 54: 49–79.

[53] Rainey, P.B. and Bailey, M.J. (1996). Physical map of the *Pseudomonas fluorescens* SBW25 chromosome. *Mol. Microbiol.* 19: 521–533.

[54] Rainey, P.B., Buckling, A., Kassen, R. and Travisano, M. (2000). The emergence and maintenance of diversity: insights from experimental bacterial populations. *Trends Ecol. Evol.* 15: 243–247.

[55] Rainey, P.B. and Travisano, M. (1998). Adaptive radiation in a heterogeneous environment. *Nature* 394: 69–72.

[56] Rendueles, O. and Ghigo, J.M. (2012). Multi-species biofilms: how to avoid unfriendly neighbors. *FEMS Microbiol. Rev.* 36: 972–989.

[57] Robertson, M., Hapca, S.M, Moshynets, O. and Spiers, A.J (2013). Air-liquid interface biofilm formation by psychrotrophic pseudomonads recovered from spoilt meat. *Antonie van Leeuwenhoek* 103: 251–259.

[58] Römling, U. (2005). Characterization of the rdar morphotype, a multicellular behaviour in *Enterobacteriaceae*. *Cell. Mol. Life Sci.* 62: 1234–1246.

[59] Sachs, J.L. and Hollowell, A.C. (2012). The origins of cooperative bacterial communities. *mBio* 3: e00099-12.

[60] Saxena, I.M., Kudlicka, K., Okuda, K. and Brown, R.M. (1994). Characterization of genes in the cellulose-synthesizing operon (*acs* operon) of *Acetobacter xylinum*: implications for cellulose crystallization. *J. Bacteriol.* 176: 5735–5752.

[61] Schirmer, T. and Jenal, U. (2009). Structural and mechanistic determinants of c-di-GMP signaling. *Nature Rev. Microbiol.* 7: 724–735.

[62] Schrecker, S.T. and Gostomski, P.A. (2005). Determining the water holding capacity of microbial cellulose. *Biotechnol. Lett.* 27: 1435–1438.

[63] Silby, M.W., Cerdeño-Tárraga, A.M., Vernikos, G., Giddens, S.R., Jackson, R., Preston, G., Zhang, X-X., Godfrey, S., Spiers, A.J., Harris, S., Challis, G.L., Morningstar, A., Harris, D., Seeger, K., Murphy, L., Rutter, S., Squares, R., Quail, M.A., Saunders, E., Anderson, I., Mavromat, K., Brettin, T.S., Bentley, S., Thomas, C.M., Parkhill, J., Levy, S.B., Rainey, P.B. and Thomson, N.R. (2009). Genomic and

functional analyses of diversity and plant interactions of *Pseudomonas fluorescens*. *Genome Biol.* 10: R51.

[64] Solomon, E.B., Niemira, B.A., Sapers, G.M. and Annous, B.A. (2005). Biofilm formation, cellulose production, and curli biosynthesis by Salmonella originating from produce, animal, and clinical sources. *J. Food Prot.* 68: 906-912.

[65] Spiers, A.J. (2007). Wrinkly-Spreader fitness in the two-dimensional agar plate microcosm: maladaptation, compensation and ecological success. *PLoS ONE* 2: e740.

[66] Spiers, A.J., Arnold, D.L., Moon, C.D. and Timms-Wilson, T.M. (2006). A survey of A-L biofilm formation and cellulose expression amongst soil and plant-associated *Pseudomonas* isolates. In: Microbial Ecology of Aerial Plant Surfaces. Bailey, M.J., Lilley, A.K., Timms-Wilson, T.M. and Spencer-Phillips, P.T.N. (Eds.). CABI, Wallingford.

[67] Spiers, A.J., Bohannon, J., Gehrig, S.M. and Rainey, P.B. (2003). Biofilm formation at the air–liquid interface by the *Pseudomonas fluorescens* SBW25 Wrinkly Spreader requires an acetylated form of cellulose. *Mol. Microbiol.* 50: 15–27.

[68] Spiers, A.J., Deeni, Y.Y., Folorunso, A.O., Koza, A., Moshynets, O. and Zawadzki, K. (2013). Cellulose expression in *Pseudomonas fluorescens* SBW25 and other environmental pseudomonads. In: Cellulose. Van De Ven, T.G.M. and Trdine, J. (Eds.). InTech, Rijeka.

[69] Spiers, A.J., Kahn, S.G., Bohannon, J., Travisano, M. and Rainey, P.B. (2002). Adaptive divergence in experimental populations of *Pseudomonas fluorescens*. I. Genetic and phenotypic bases of Wrinkly Spreader fitness. *Genetics* 161: 33–46.

[70] Spiers, A.J. and Rainey, P.B. (2005). The *Pseudomonas fluorescens* SBW25 Wrinkly Spreader biofilm requires attachment factor, cellulose fibre and LPS interactions to maintain strength and integrity. *Microbiology* 151: 2829–2839.

[71] Sullivan, S.M. and Maddock, J.R. (2000). Bacterial division: finding the dividing line. *Curr. Biol.* 10: R249–252.

[72] Travisano, M. (2001). Evolution: Towards a genetical theory of adaptation. *Curr. Biol.* 11: R440–R442.

[73] Travisano, M. and Rainey, P.B. (2000). Studies of Adaptive Radiation Using Model Microbial Systems. *Amer. Nat.* 156: S35–S44.

[74] Ude, S., Arnold, D.L., Moon, C.D., Timms-Wilson, T. and Spiers, A.J. (2006). Biofilm formation and cellulose expression among diverse environmental *Pseudomonas* isolates. *Environ. Microbiol.* 8: 1997–2011.

[75] Ude, S., Bailey, M.J., Huang, W.E. and Spiers, A.J. (2007). The environmental plasmid pQBR103 alters the single-cell Raman spectral profile of *Pseudomonas fluorescens* SBW25. *Microb. Ecol.* 53: 494–497.

[76] Van Valen, L. (1973). A New Evolutionary Law. *Evol. Theory* 1: 1–30.

[77] Verschuren, P.G., Cardona, T.D., Nout, M.H.R., De Gooijer, K.D. and Van Den Heuvel, J.C. (2000). Location and limitation of cellulose production by *Acetobacter xylinum* established from oxygen profiles. *J. Biosc. Bioeng.* 89: 414–419.

[78] Webb, J.S., Givskovy, M. and Kjelleberg, S. (2003). Bacterial biofilms: prokaryotic adventures in multicellularity. *Curr. Opin. Microbiol.* 6: 578–585.

[79] West, S.A., Griffin, A.S., Gardner, A. and Diggle, S.P. (2006). Social evolution theory for microorganisms. *Nature Rev. Microbiol.* 4: 597–607.

[80] Winogradsky, S. (1887). Ueber Schwefelbacterien. *Bot. Zeit.* 45: 489–610.

[81] Zogaj, X., Bokranz, W., Nimtz, M. and Römling, U. (2003). Production of cellulose and curli fimbriae by members of the family *Enterobacteriaceae* isolated from the human gastrointestinal tract. *Infect. Immun.* 71: 4151–4158.

In: Microcosms
Editor: Christopher C. Harris

ISBN: 978-1-62618-661-3

Chapter 6

Ecology and Environmental Side-Effects of Pesticides in Tropical Microcosms

M. A. Daam[1*], R. Geber-Corrêa[2] and A. V. Waichman[2]
[1]Instituto Superior de Agronomia, Technical University of Lisbon, Tapada da Ajuda, Lisbon, Portugal
[2]Universidade Federal do Amazonas, Faculdade de Ciências Agrárias, Manaus, Amazonas, Brazil

Abstract

It has often been discussed that research into the environmental fate and effects of pesticides has been almost exclusively conducted under non-tropical conditions. Furthermore, those studies conducted in tropical regions have focused almost exclusively on single species laboratory tests and very few microcosm studies under tropical settings are currently available. Given the differences in ecology and climatic conditions between tropical and non-tropical conditions, differences in pesticide fate, direct and indirect effects and recovery potential of impaired ecosystems may be anticipated. In the present chapter, findings from studies conducted in tropical Thailand are included to illustrate these possible differences. Indications for the methodology and research needs of future tropical microcosm studies are discussed.

Introduction

As a result of a shift from traditional subsistence farming toward intensive-crop farming, pesticide use in tropical countries has increased considerably since the Green Revolution in the late 1960s (e.g., Satapornvanit et al. 2004; SINDAG, 2009). Capable of growing two to three crops a year, tropical countries are becoming the future "breadbaskets" for the world, exporting agricultural produce to regions having colder climates and shorter growing seasons

---

* Corresponding author: Michiel A. Daam, Instituto Superior de Agronomia / Technical University of Lisbon, Tapada da Ajuda, 1349-017 Lisbon, Portugal, Telephone: +351 21 365 3224 (direct), Email: mdaam@isa.utl.pt / michieldaam@gmail.com.

(Ecobichon, 2001; Satapornvanit et al., 2004). This also implies that pesticides are applied throughout a larger part of the year, which may impede ecosystems to recover from pesticide stress. Daam and Van den Brink (2010a) recently reported that tropical freshwater ecosystems surrounding agricultural fields may be especially at risk to pesticide stress due to i) high levels of spray drift since applications are often made in close proximity to such water bodies; ii) large amounts of runoff resulting from intensive irrigation practices and episodes of heavy tropical precipitation; iii) greater rate and frequency of pesticide applications (including misuse practices); iv) dangerous transportation and storage conditions; v) the use of cheaper and more hazardous pesticides; and vi) washing of application equipment in water bodies.

Despite this great increase in pesticide use in the tropics over the past decades, ecotoxicological research into their environmental fate and side effects has focused almost exclusively on temperate countries (Daam and Van den Brink, 2010a). Subsequently, risk assessments in tropical countries often rely on temperate toxicity data, although it may be debatable whether the fate and effects of chemicals are comparable in geographically distinct ecosystems. In addition, the few ecotoxicological studies carried out in the tropics applied techniques and procedures developed for temperate environments, even though differences in physical, chemical and biological parameters in the tropics may dictate that new approaches have to be developed (Lacher and Goldstein, 1997).

Studies into comparisons of temperate and tropical freshwater responses towards pesticides are almost exclusively based on toxicity data from single species laboratory tests using temperate and tropical taxa which are compared using the species sensitivity distribution (SSD) concept (e.g., Kwok et al., 2007; Rico et al., 2010). Studies at higher-tier levels have hardly been carried out in the tropics, implying that i) fate, ii) direct effects on community level, iii) indirect effects, and iv) recovery potential are currently poorly understood (Daam and Van den Brink, 2011).

To contribute to this lack in knowledge, outdoor microcosm experiments were conducted with different pesticides in Thailand (the insecticide chlorpyrifos, the fungicide carbendazim, and the herbicide linuron). Experimental set-up and methodologies used for endpoint measurements were based on those developed for and proven adequate in temperate microcosm studies. In the present chapter, study findings from these studies are presented to discuss i) the ecology of tropical (model) ecosystems, ii) pesticide fate and sensitivity of tropical freshwater life to pesticide stress in microcosms, and iii) Indications for the methodology and research needs of future tropical microcosm studies.

Methodology of the Microcosm Studies

Two different test systems were used to conduct the microcosm experiments at the Asian Institute of Technology (AIT, Bangkok, Thailand). Rectangular microcosms (length: 1 m; width: 1 m; height: 1.15 m) were used to evaluate the fate and effects of single applications of the insecticide chlorpyrifos and the fungicide carbendazim (Figure 1). The impact of multiple chlorpyrifos applications and a single application of the herbicide linuron were evaluated using circular microcosms (diameter: 0.75 m; height: 0.65 m; Fig. 1). Both test systems were constructed from concrete and were freshly coated with nontoxic epoxy paint before each experiment to prevent influences from chemicals diffusing from the concrete itself or still present from earlier experiments. The rectangular tanks were filled with a 10-cm layer of sediment and a 1-m water column (corresponding with 1000 L), taken from the canal surrounding AIT. The circular tanks were devoid of sediment and contained a water column of 0.55 m (water volume approximately 250 L). In all experiments, the canal water was filtered through a net (mesh size, 0.1 mm) before it was added to the microcosms to prevent fish and prawns from entering the experimental systems. The microcosms were intended to model the community of Thai farm canals.

Figure 1. Microcosms used for the experiments in Thailand: rectangular microcosms (A; length: 1 m; width: 1 m; height: 1.15 m) and the circular microcosms (B; diameter: 0.75 m; height: 0.65 m).

To enable sampling of macroinvertebrate and periphyton communities, artificial substrates were constructed and deployed in the microcosms. Macroinvertebrate substrates were similar to the design described by Brock et al. (1992) and consisted of pebble stone baskets that were placed on top of the sediment. Periphyton was sampled from microscopic slides positioned in Perspex frames that were suspended 2 weeks before pesticide application at approximately 10 cm below the water surface. Litter bags filled with *Musa* (banana) leaves were used to study decomposition of particulate organic matter after incubation periods of 2, 4, and 8 weeks. Other endpoints were determined by applying methods in accordance with those used in temperate studies and included pesticide analysis in water column (and sediment in the experiment evaluating a single chlorpyrifos application), physico-chemical parameters (dissolved oxygen, pH, electrical conductivity, temperature, alkalinity, ammonia, nitrate, and ortho-phosphate), chlorophyll-a (phytoplankton and periphyton), and community structure of phytoplankton and zooplankton. For more details, the reader is referred to Daam et al. (2008, 2009a, b).

Ecology of Tropical (Model) Ecosystems

Species Diversity

Species diversity has often been discussed to increase towards the tropics, indicating that the number of species that could potentially be affected by pollutants is also greater (Gaston et al., 1995; Mares, 1997; Lacher and Goldstein, 1997; Kwok et al., 2007). Species diversity, however, appears not to be greater in the tropics for all taxonomic groups (Daam and Van den Brink, 2010b). For example, although freshwater fish species richness increases towards the equator (e.g., Lévêque et al., 2008), freshwater plankton communities do not show a marked latitudinal trend in species diversity, and there may even be a minor trend towards a lower diversity at low latitudes (Kalff and Watson, 1986; Lewis, 1987; Fernando, 2002). Also within taxonomic groups, diversity may be lower or higher in tropical freshwaters as compared to their temperate counterparts (Daam and Van den Brink, 2010b). For example, the insect families Chironomidae and Culicidae, both belonging to the insect order Diptera, have respectively greater and lesser species diversity in the Palearctic and Nearctic regions compared to the Neotropical, Afrotropical and Oriental regions (Balian et al., 2008). A comparison of species diversity between the temperate and tropical zone has often been reported to be biased by the fact that taxonomic expertise and research efforts have centered on temperate regions (Mares, 1997; Dudgeon, 2000; Sarma et al., 2005; Balian et al., 2008). This lacune in knowledge is due, in part, to a lack of basic research, but is also a reflection of a scattered, highly fragmentated literature, some of which is inaccessible (Dudgeon, 2000).

Ecosystem Structure

Given the enormous natural variability in the structure and function of freshwater communities, it is reasonable to question the geographical extrapolation of model ecosystem experiments. Since most model ecosystems enclose parts of – or have been seeded with components of – natural communities, the geographical location of micro- and mesocosms will determine their species composition and hence potentially their sensitivity (Brock et al., 2008).

Various differences in species composition of the various taxonomic groups present in ecosystems have indeed been reported. For example, due to the year-round predation by the great diversity of fish and invertebrate predators, and increased metabolic costs with increasing temperatures, large cladocerans like *Daphnia* have been reported to be practically absent in the tropics (e.g., Duncan, 1984; Fernando, 1994; Dumont, 1994). Tropical water bodies typically contain cladocerans from the genera *Moina*, *Ceriodaphnia*, *Macrothrix* and *Diaphanosoma* (Sarma et al., 2005). The high tropical temperature further determines the distribution in tropical water bodies of mainly warmwater adapted rotifer species of which *Brachionus caudatus*, *B. plicatilis*, *B. falcatus*, *B. calyciflorus*, *Filinia opoliensis*, *Hexarthra mira*, *Keratella tropica*, *Anuraeopsis fissa*, and *Asplanchna brightwelli* are the most dominant (Kutikova, 2002). Taxa belonging to the genus *Lecane* have also been reported to be well-represented in tropical freshwaters (Dussart et al., 1984; Segers, 2008). Furthermore, Ostracoda are rare in temperate freshwaters but quite common in tropical waters (Dussart et al., 1984; Victor, 2002). In line with this, the main cladoceran taxa that were encountered in the Thai microcosm studies were *Ceriodaphnia cornuta* and *Moina micrura*, species that have indeed been reported as eurytopic in tropical Asian freshwaters (Sarma et al., 2005). Rotifers (mainly taxa belonging to the genera *Brachionus*, *Lecane*, and *Filinia*; Daam and Van den Brink, 2010b) and copepods (especially immature stages, i.e. Nauplii) had the greatest abundances in terms of numbers per litre, which has also been discussed to be typical for tropical freshwaters (e.g. Fernando, 1994, 2002).

Regarding phytoplankton, Kalff and Watson (1986) concluded that the proportion of phytoplankton in the major taxonomic and size divisions is not so much a function of latitude as of nutrient levels and mixing regime after comparing the phytoplankton biomass and community structure of tropical and temperate lakes. For example, these authors discussed that the relative importance of diatoms, with their heavy frustules and rapid sinking rate, is more related to levels of silica, phosphorous and mixing regime than to latitude (Kalff and Watson, 1986).

The diversity and species composition of aquatic higher plants, however, has been demonstrated to differ over different zoogeographical regions. For example, after reviewing the global diversity of freshwater macrophytes, Chambers et al. (2008) concluded that macrophyte species diversity is highest in the Neotropics, intermediate in the Oriental, Afrotropical, and Nearctic, and relatively low in the Australasian and the Palaearctic regions. The higher number of species in the Neotropics is in great part due to the large contribution from the Podostemaceae (188 species) compared to other regions. Plants in this family are confined to fast-flowing waters, mainly in the tropics, and many species have narrow distributions, such as a single watershed. For example, the genus *Apinagia* is found only in South America (Chambers et al., 2008). In addition, about 39% of the c. 412 genera containing aquatic vascular macrophytes are endemic to a single biogeographic region, with 61–64% of all aquatic vascular plant species found in the Afrotropics and Neotropics being endemic to those regions (Chambers et al., 2008). This macrophyte diversity and high degree of endemism implies that the existence of taxa with greater sensitivities than the standard macrophyte test species commonly used in temperate regions (i.e. *Lemna* spp and, to a lesser extent and more recently, *Myriophyllum* spp) is also greater in the latter regions.

No macrophytes, however, were included in the Thai microcosms since the drainage canals in Thai farms were noted to be mostly devoid of macrophytes (personal observation). Daam et al. (2009a) distinguished three possible underlying reasons for this i) macrophytes have been discussed to be susceptible to tropical hydrological regimes that produce marked changes in water level (Talling and Lemoalle, 1998); ii) small boats, used for irrigation and pesticide application purposes, are often pushed or pulled through the farm canals, by walking through the canal which disturbs the sediment and hampers macrophyte growth; and iii) the invasion of the origin South American apple snails in Thailand has been associated with an almost complete absence of aquatic plants, with high nutrient concentrations and with high phytoplankton biomass (Carlsson et al., 2004).

Aquatic macrophytes play an important role in the structure and function of aquatic ecosystems, as they provide habitat and food to many organisms (Chambers et al., 2008). Compared with ecosystems dominated by macrophytes, plankton-dominated model ecosystems are usually characterized by a higher proportion of short-lived species (phytoplankton, zooplankton), lower biomass but higher turnover rates, and a less diverse macroinvertebrate community (Brock and Budde, 1994). In line with this, total numbers and diversity of the macroinvertebrate communities in the Thai microcosms were low, which hampered evaluating both the sensitivity and recovery potential of macroinvertebrate populations and communities after pesticide stress (Daam et al., 2008, 2009b).

Tropical Limnology

Basis ecological processes are similar in tropical and temperate lakes including grazing, competition, predation and abiotic adaptation; the major difference being the greater speed of the processes in the warm tropics (Nilssen, 1984; Lewis, 1987; Kutikova, 2002). However, temperate limnology is not immediately applicable to tropical countries because the nature of the biota and characteristic pathways and biological processes differ to a large extent (Schiemer, 1995). According to Fernando (1994), the major difference between fish and zooplankton in tropical and temperate lakes is the predominance of Rotifera and herbivorous fish in tropical lakes versus crustacea and non-herbivorous fish in temperate lakes (Figure 2). The relative biomass of zooplankton/phytoplankton is low in the tropics and therefore, unlike in temperate lakes, zooplankton does not control phytoplankton biomass in the tropics (Piyasiri and Perera, 2001).

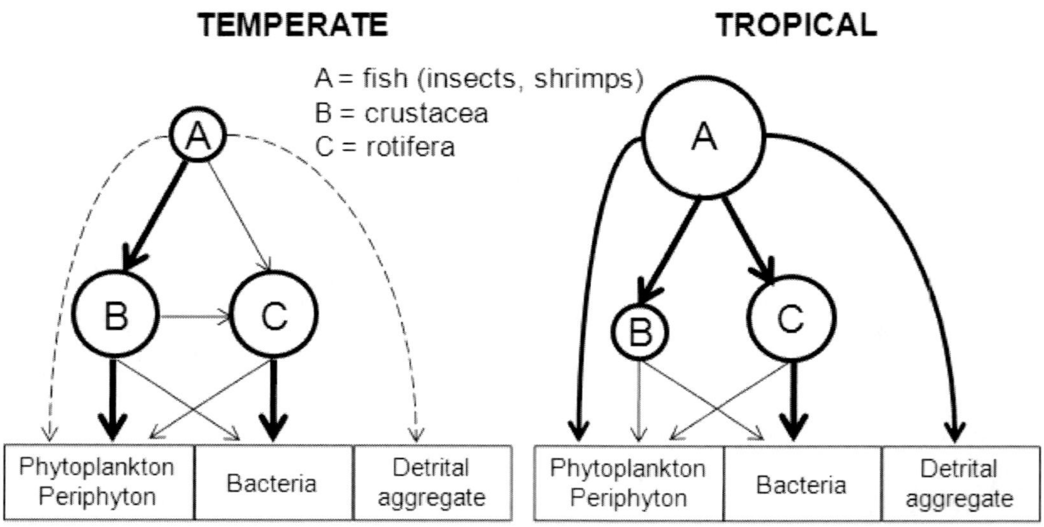

Figure 2. Main freshwater ecosystem components and their food interrelationships in temperate versus tropical freshwaters (Modified from Fernando, 1994). Sizes of the circles indicate the relative size of the ecosystem components, the thickness of the arrow the strength of the interaction between them, while its direction points to the consumer / food relationship.

Studies into the community composition and succession of periphyton in tropical freshwaters have been minimal, although periphyton serves as food for many invertebrates and fishes (Talling and Lemoalle, 1998; Pekthong and Peerepornpisal, 2001). Since effects on periphyton could be expected in the Thai microcosm study evaluating the herbicide linuron, the periphyton communities were studied in this experiment using glass slides that served as artificial substratum. The slides were positioned in a glass frame that was suspended 2 weeks before application at approximately 10 cm below the water surface. At 2-week intervals (up to 10 weeks post incubation), the periphyton biomass of five slides was collected by brushing the slides visually clean, after which the sample was preserved with lugol and taxa identified using an inverted microscope.

In temperate freshwaters, the succession of periphytic algae on a pristine surface is from small, flat cells with a large surface area attached to the substrate like diatoms to standing or stalked forms and eventually filamentous forms (Brönmark and Hansson, 2005). In line with this, dominance of the filamentous colony-forming cyanobacterium *Chamaesiphon* sp. was only reached after four weeks. Two possible causes for the complete disappearance of this species eight weeks after incubation of the substrates are an increase in the number of snails grazing on the substrates and the occurrence of two tropical rainshowers in this period (Daam and Van den Brink, 2010b).

Fate of Pesticides in Tropical Freshwaters and Their Environmental Side-Effects

Fate

Due to differences in climate parameters, dissipation of chemicals from the aquatic environment have been suggested to be generally higher in tropical regions compared to temperate regions (e.g. Bourdeau et al. 1989; Magallona 1994; Racke 2003). To determine whether fate may be expected to differ between temperate and tropical settings, Sanchez-Bayo and Hyne (2011) recently used data from the literature and modelling outputs (Fugacity modelling level 3) based on the physicochemical properties of the compounds. They concluded that differences regarding the faster dissipation rates in warm and wet environments may translate into shorter exposures and higher rates of recovery of the organisms affected in tropical regions. However, the high precipitation in the tropics will undoubtedly increase the risk of runoff for most compounds and foster the leaching potential of certain chemicals. At the same time, other features specific to particular tropical regions may accentuate the ecological risks: poor soils may lead to a higher mobility of residues into the water phase, whether surface water or groundwater, and acidic soils contribute to this mobility, thus exacerbating the exposure of the soil microfauna and microbial communities to herbicides and other polar compounds. In summary, the balance between positive and negative risks appears to depend on the characteristics of the specific soils in each region, rather than on climatic characteristics of the tropics (Sanchez-Bayo and Hyne, 2011), a conclusion similar to that reached by Daam and Van den Brink (2010a) after reviewing differences in climate-related parameters between temperate and tropical conditions.

Model ecosystem studies evaluating the fate of pesticides under tropical conditions also do not consistently demonstrate faster disappearances of pesticides than in studies performed in the temperate zone. Dissipation rates of alachlor, atrazine, endosulfan, metalochlor, profenofos, simazine and trifluralin in outdoor aquatic microcosms set up in Brazil were not distinctively greater in comparison to temperate (semi-) field or laboratory studies (Laabs et al. 2007). Lahr et al. (2000) concluded that, given the results from laboratory studies, bendiocarb and fenitrothion were surprisingly persistent in tropical experimental ponds in the Sahel. Regarding the studies conducted in Thailand, disappearance rates of the insecticide chlorpyrifos were also similar to those found by studies in temperate regions (Daam et al., 2008). Carbendazim disappeared slightly faster from the water column than reported in temperate model ecosystem studies, which was explained by a possibly greater microbial breakdown of the fungicide under the tropical conditions tested (Daam et al., 2009b).

Environmental Side-Effects

Studies aiming at a validation of the protective value of temperate toxicity threshold values for tropical freshwaters have focused mainly on a species (assemblage)-level based approach by comparing sensitivities of species between temperate and tropical freshwaters using Species Sensitivity Distributions (SSDs). The authors of these studies all concluded that there is no consistent difference in sensitivity between temperate and tropical species (e.g. Dyer et al. 1997; Maltby et al. 2005; Kwok et al., 2007; Rico et al., 2010). Based on their analysis, Kwok et al. (2007) recommended the use of an extrapolation factor of 10 for coverage of 95% of the chemicals with a 90% protection level if the water quality standard is primarily based on temperate species and a priori knowledge of the sensitivity of tropical species is very limited or not available.

Model ecosystem studies evaluating the effects of a single application of the insecticide chlorpyrifos performed in different climates (temperate, Mediterranean, and tropical) all resulted in a NOECecosystem of 0.1 µg/l (Table 1). NOECecosystem values for the fungicide carbendazim and the herbicide linuron calculated in the microcosm studies conducted in Thailand also appeared to be similar or higher compared to those reported in temperate studies (Daam et al. 2010a).

Table 1. No observed effect concentration (NOEC; in µg/L), lowest observed effect concentration (LOEC; in µg/L), and weeks to recovery of the zooplankton community at LOEC-level reported in model ecosystem studies evaluating short-term exposure (single or pulsed applications) of the insecticide chlorpyrifos under temperate, mediterranean and tropical conditions

Location	Type of system	NOEC	LOEC	Weeks to recovery	Reference
Temperate					
The Netherlands	Lentic, outdoor, experimental ditches	0.1	0.9	4	Van den Brink et al. (1996)
Kansas, USA	Lentic, outdoor, microcosms	0.1	0.3 - 1	< 4	Biever et al. (1994)
The Netherlands	Lentic, indoor, microcosms				
	Mesotrophic; cool (16–18 °C)	0.1	1	3	Van Wijngaarden et al. (2005)
	Mesotrophic; warm (24–28 °C)	0.1	1	5	Van Wijngaarden et al. (2005)
	Eutrophic; warm (25–28 °C)	0.1	1	> 5	Van Wijngaarden et al. (2005)
Mediterranean					
Spain	Lentic, outdoor, mesocosms	0.1	1	8 – 11	López-Mancisidor et al. (2008)
Tropical					
Thailand	Lentic, outdoor, microcosms	0.1	1	9	Daam et al. (2008)

Evidently, additional tropical model ecosystem studies evaluating single pesticide concentration ranges are required to evaluate whether this is also valid for a wider array of compounds and on a larger tropical geographical scale. However, there is little doubt that structurally different ecosystems respond differently to toxic stress when exposed to concentrations well above threshold concentrations for direct toxic effects (Brock et al., 2008). Indeed, despite the agreement in NOECecosystem values for chlorpyrifos (Table 1), algae blooms as an indirect effect due to the death of various zooplankters was more pronounced in the Thai microcosm study and, hence, recovery of the zooplankton community took longer than that reported in temperate studies (Daam et al., 2008). Interestingly, Van Wijngaarden et al. (2005; indoor) and López-Mancisidor et al. (2008; outdoor) also reported a greater time-to-recovery for cladoceran populations affected by chlorpyrifos underwarmer (Mediterranean) conditions as compared to temperate conditions. Although only based on Corixidae, macroinvertebrates also appeared to recover slightly slower under tropical as compared to temperate settings (Daam et al., 2008). In line with this, Lahr and co-workers reported that the fairy shrimp *Streptocephalus sudanicus* differed highly in their response to insecticides compared to Daphnia magna and recovered extremely slowly compared to many other invertebrate groups (Lahr et al., 2000, 2001). Contrarily, the recovery potential of affected endpoints (community metabolism and plankton communities) after linuron application appeared higher for tropical ecosystems compared to their temperate counterparts (Daam et al., 2009a). The need for additional studies into the recovery potential of pesticide stressed tropical freshwaters may be further emphasized by year-round use of pesticides in the tropics versus temperate agricultural practices with only one distinct growing season, and associated period of pesticide use, per year (Daam and Van den Brink, 2010a).

Indications for Future Model Ecosystem Studies in the Tropics

Recently, Daam and Van den Brink (2011) scrutinized various improvements and important aspects that should be considered when conducting model ecosystem experiments in the tropics, as well as various indications for future research, which are summarized below (for more information: see Daam and Van den Brink, 2011):

- Given the discussed importance of fish and shrimps in structuring tropical freshwaters (Figure 2), it would be interesting to carry out model ecosystem studies that include indigenous representatives of these organisms to study trophic interrelationships on a semi-field level. Chemicals like pesticides could be used to knock down a defined (i.e., depending on the mode of action of the chemical) part of the foodweb so ecosystem interactions can be studied.
- Macroinvertebrates in Thai canals were noted to be concentrated in the roots of floating plants. Since no floating plants were included in the model ecosystem experiments (since they could cover the entire water surface), may explain the generally low numbers and diversity of macroinvertebrates in the cosms. Hence, developing substrates that mimic floating plants as an addition to the pebble stone baskets seems meaningful.
- Expertise and knowledge on identification of tropical macroinvertebrates is limited, and hence, studies into identification of tropical macroinvertebrates are urgently needed.

- Additional studies into the recovery potential of pesticide stressed tropical freshwaters are needed.
- Future outdoor model ecosystems studies evaluating herbicides under a tropical monsoon climate (e.g., Thailand) should be carried out in the end of the cold season or the beginning of the hot season, since the climatic conditions in the rainy season appear to favourite cyanobacteria blooms.
- Experiments into the optimization of nutrient additions of nitrogen and phosphorous to model ecosystems may be needed. In addition, the possible need for additions of chemicals other than nitrogen and phosphorous (e.g. silica and/or carbon) should be evaluated.
- Sediment should be included in future model ecosystem studies to avoid a drop in alkalinity levels over the course of the experiment and to ensure that exposure conditions are closer to real-world situations.
- Additional tropical model ecosystem studies evaluating single pesticide concentration ranges are required to evaluate whether the consistency in NOEC values between temperate and tropical studies is also valid for a wider array of compounds and on a larger tropical geographical scale.
- Additional experimental research is required evaluating repeated applications of pesticides with different degradation rates and application intervals relevant for local agricultural practices to provide a better understanding of pesticide freshwater ecotoxicology in tropical countries like Thailand.
- Given the reported widespread contamination of pesticides over tropical watersheds and the variety of aquatic ecosystems that exist in the tropics, a greater variety of tropical model ecosystems has to be developed. These systems should account for the (physical, chemical and biological) variety in aquatic ecosystems potentially at risk as to improve our understanding of pesticide risk in tropical regions on a watershed level.

Acknowledgment

The studies described in this chapter that were conducted in Thailand were funded by the Portuguese government through a research grant project of FCT (*Fundação para a Ciência e Tecnologia*) with reference PTDC/AGR-AAM/105459/2008, the European Commission's INCO: International Scientific Cooperation Projects (MAMAS project, contract number ICA4-2000-10247).

References

Balian EV, Segers H, Lévêque C, Martens K (2008). The freshwater animal diversity assessment: an overview of the results. *Hydrobiologia* 595: 627-637.

Biever RC, Giddings JM, Kiamos M, Annunziato MF, Meyerhoff R, Racke K (1994). *Effects of chlorpyrifos on aquatic microcosms over a range of off-target drift exposure levels.* In:

Proceedings, Brighton crop protection conference-pests and diseases, Brighton, UK, November, 21–24. pp 1367–1372.

Bourdeau P, Haines JA, Klein W, Krishna Murti CR, eds (1989) *Ecotoxicology and climate: with special reference to hot and cold climates*. Wiley, Chichester, UK.

Brock TCM, Van den Bogaert M, Bos AR, Van Breukelen SWF, Reiche R, Terwoert J, Suykerbuyk REM, Roijackers RMM (1992). Fate and effects of the insecticide Dursban 4E in indoor *Elodea*-dominated and macrophyte-free model ecosystems: II. Secondary effects on community structure. *Arch Environ Con Tox* 23: 391-409.

Brock TCM, Budde BJ (1994). On the choice of structural parameters and endpoints to indicate responses of freshwater ecosystems to pesticide stress. In: Hill IR, Heimbach F, Leeuwangh P, Matthiessen P (eds). *Freshwater Field Tests for Hazard Assessment of Chemicals*, Lewis Publishers, Boca Raton, FL, USA, pp 19-56.

Brock TCM, Maltby L, Hickey CH, Chapman J, Solomon KR (2008). Spatial extrapolation in ecological effect management of chemicals. In: Solomon KR, Brock TCM, De Zwart D, Dyer SD, Posthuma L, Richards SM, Sanderson H, Sibley PK, Van den Brink PJ (eds). *Extrapolation Practice for Ecotoxicological Effect Characterization of Chemicals*, SETAC Europe Press, Brussels, Belgium, pp 223-256.

Brönmark C, Hansson LA (2005). *The biology of lakes and ponds* (2nd edition). Oxford University Press, UK.

Carlsson NOL, Brönmark C, Hansson L.A (2004). Invading herbivory: the golden apple snail alters ecosystem functioning in Asian wetlands. *Ecology* 85: 1575-1580.

Chambers PA, Lacoul P, Murphy KJ, Thomaz SM (2008). Global diversity of aquatic macrophytes in freshwater. *Hydrobiologia* 595: 9-26.

Daam MA, Crum SJH, Van den Brink PJ, Nogueira AJA (2008). Fate and effects of the insecticide chlorpyrifos in outdoor plankton-dominated microcosms in Thailand. *Environ Toxicol Chem* 27: 2530-2538.

Daam MA, Van den Brink PJ, Nogueira AJA (2009a). Comparison of fate and ecological effects of the herbicide linuron in freshwater model ecosystems between tropical and temperate regions. *Ecotoxic Environ Safe* 72: 424-433.

Daam MA, Satapornvanit K, Van den Brink PJ, Nogueira AJA (2009b). Sensitivity of macroinvertebrates to carbendazim under semi-field conditions in Thailand: Implications for the use of temperate toxicity data in a tropical risk assessment of fungicides. *Chemosphere* 74: 1187-1194.

Daam MA, Van den Brink, PJ (2010a). Implications of differences between temperate and tropical freshwater ecosystems for the ecological risk assessment of pesticides. *Ecotoxicology* 19: 24-37.

Daam MA, Van den Brink PJ (2010b). Biodiversity, succession and seasonality of tropical freshwater plankton communities under semi-field conditions in Thailand. In: De Carlo F, Bassano A (eds.). *Freshwater Ecosystems and Aquaculture Research*, Nova Science Publishers, New York, USA, pp 225-248.

Daam MA, Van den Brink PJ (2011). Conducting model ecosystem studies in tropical climate zones: lessons learned from Thailand and way forward. *Environ Pollut* 159: 940-946.

Dudgeon D (2000). The ecology of tropical Asian rivers and streams in relation to biodiversity conservation. *Annu Rev Ecol Syst* 31: 239-263.

Dumont HJ (1994). On the diversity of the Cladocera in the tropics. *Hydrobiologia* 272: 27-38.

Duncan A (1984). Assessment of factors influencing the composition, body size and turnover rate of zooplankton in Parakrama Samudra, an irrigation reservoir in Sri Lanka. *Hydrobiologia* 113: 201-215.

Dussart BH, Fernando CH, Matsumura-Tundisi T, Shiel RJ (1984). A review of systematics, distribution and ecology of tropical freshwater zooplankton. *Hydrobiologia* 113: 77-91.

Dyer SD, Belanger SE, Carr GJ (1997). An initial evaluation of the use of Euro/North American fish species for tropical effects assessments. *Chemosphere* 35: 2767–2781.

Ecobichon DJ (2001). Pesticide use in developing countries. *Toxicology* 160: 27-33.

Fernando CH (1994). Zooplankton, fish and fisheries in tropical freshwaters. *Hydrobiologia* 272: 105-123.

Fernando CH (2002). Zooplankton and tropical freshwater fisheries. In: Fernando CH (ed). A guide to tropical freshwater zooplankton, Backhuys Publishers, Leiden, The Netherlands, pp 255-280.

Gaston KJ, Williams PH, Eggleton P, Humphries CJ (1995). Large scale patterns of biodiversity: spatial variation in family richness. *P Roy Soc Lond B Biol* 260: 149-154.

Kalff J, Watson S (1986). Phytoplankton and its dynamics in two tropical lakes: a tropical and temperate zone comparison. *Hydrobiologia* 138: 161-176.

Kutikova LA (2002). Rotifera. In: Fernando CH, ed. *A guide to tropical freshwater zooplankton*. Backhuys Publishers, Leiden, the Netherlands. pp: 23-68.

Kwok KWH, Leung KMY, Chu VKH, Lam PKS, Morritt D, Maltby L, Brock TCM, Van den Brink PJ, Warne MStJ, Crane M (2007). Comparison of tropical and temperate freshwater species sensitivities to chemicals: implications for deriving safe extrapolation factors. *Integrated Environ Assessment Manage* 3: 49-67.

Laabs V, Wehrhan A, Pinto A, Dores E, Amelung W (2007). Pesticide fate in tropical wetlands of Brazil: an aquatic microcosm study under semi-field conditions. *Chemosphere* 67: 975–989.

Lacher TE, Jr, Goldstein MI (1997). Tropical Ecotoxicology: status and needs. *Environ Toxicol Chem* 16: 100-111.

Lahr J, Diallo AO, Gadji B, Diouf PS, Bedaux JJM, Badji A, Ndour KB, Andreasen JE, Van Straalen NM (2000). Ecological effects of experimental insecticide applications on invertebrates in Sahelian temporary ponds. *Environ Toxicol Chem* 19: 1278–1289.

Lahr J, Badji A, Marquenie S, Schuiling E, Ndour KB, Diallo AO, Everts JW (2001). Acute toxicity of locust insecticides to two indigenous invertebrates from Sahelian temporary ponds. *Ecotox Environ Saf* 48: 66–75.

Lévêque C, Oberdorff T, Paugy D, Stiassny MLJ, Tedesco PA (2008). Global diversity of fish (Pisces) in freshwater. *Hydrobiologia* 595: 545-567.

Lewis WM Jr (1987). Tropical limnology. *Annu Rev Ecol Syst* 18: 159-184.

López-Mancisidor P, Carbonell G, Marina A, Fernández C, Tarazona JV (2008). Zooplankton community responses to chlorpyrifos in mesocosms under Mediterranean conditions. *Ecotox Environ Saf* 71: 16–25.

Magallona ED (1994). Impact of pesticides on tropical ecosystems with emphasis on rice production in the Asian countries. In: Widianarko B, Vink K, Van Straalen NM (eds). *Environmental toxicology in South East Asia*. VU University Press, Amsterdam, the Netherlands. pp. 19–31.

Maltby L, Blake N, Brock TCM, Van den Brink PJ (2005). Insecticide species sensitivity distributions: importance of test species selection and relevance to aquatic ecosystems. *Environ Toxicol Chem* 24: 379–388.

Mares MA (1997). Tropical ecotoxicology: explorers for a new era. *Environ Toxicol Chem* 16: 1-2.

Nilssen JP (1984). Tropical lakes – functional ecology and future development: The need for a process-orientated approach. *Hydrobiologia* 113: 231-242.

Pekthong T, Peerapornpisal Y (2001). Fifty one new record species of freshwater diatoms in Thailand. *Chiang Mai Journal of Science* 28: 97-112.

Piyasiri S, Perera N (2001). Role of *Oreochromis* hybrids in controlling *Microcystis aeruginosa* blooms in the Kotmale Reservoir. In: de Silva SS (ed.). *Reservoir and Culture-Based Fisheries: Biology and Management - Proceeding of an International Workshop*. Bangkok, Thailand. pp. 137-148.

Racke KD (2003*). What do we know about the fate of pesticides in tropical ecosystems? Environmental fate and effects of pesticides ACS symposium series* 853: 96–123.

Rico A, Geber-Corrêa R, Campos PS, Garcia MVB, Waichman AV, Van den Brink PJ (2010). Effect of parathion-methyl on Amazonian fish and freshwater invertebrates: a comparison of sensitivity with temperate data. *Arch Environ Con Tox* 58: 765-771.

Sanchez-Bayo F, Hynez RV (2011). Comparison of environmental risks of pesticides between tropical and nontropical regions. *Integrated Environ Assessment Manage* 7: 577–586.

Sarma SSS, Nandini S, Gulati RD (2005). Life history strategy of cladocerans: comparisons of tropical and temperate taxa. *Hydrobiologia* 542: 315-333.

Satapornvanit K, Baird DJ, Little DC, Milwain GK, Van den Brink PJ, Beltman WHJ, Nogueira AJA, Daam MA, Domingues I, Kodithuwakku SS, Perera MWP, Yakupitiyage A, Sureshkumar SN, Taylor GJ (2004). Risks of pesticide use in aquatic ecosystems adjacent to mixed vegetable and monocrop fruit growing areas in Thailand. *Aust J Ecotoxicol* 10: 85-95.

Schiemer F (1995). Bottom-up vs. top-down control in tropical reservoir management. In: Timotius KH, Golthenboth F, ed. *Tropical Limnology 1*, Satya Wacanan Christian University, Indonesia. pp. 57–67.

Segers H (2008). Global diversity of rotifers (Rotifera) in freshwater. *Hydrobiologia* 595: 49-59.

[SINDAG] Sindicato Nacional de Industria de Produtos para Defesa Agrícola (2009). Brasil lidera uso mundial de agotóxicos. [online] 7 August. Available at *http://www.sindag.com.br/noticia.php?News_ID=1755.* [Accessed on 30 November 2010]

Talling JF, Lemoalle J (1998). *Ecological Dynamics of Tropical Inland Waters*. Cambridge University Press, UK.

Van den Brink PJ, Van Wijngaarden RPA, Lucassen WGH, Brock TCM, Leeuwangh P (1996). Effects of the insecticide Dursban 4E (active ingredient chlorpyrifos) in outdoor experimental ditches: II. Invertebrate community responses and recovery. *Environ Toxicol Chem* 15: 1143–1153.

Van Wijngaarden RPA, Brock TCM, Douglas MT (2005). Effects of chlorpyrifos in freshwater model ecosystems: the influence of experimental conditions on ecotoxicological thresholds. *Pest Manage Sci* 61: 923–935.

Victor R (2002). Ostracoda. In: Fernando CH, ed. A Guide to Tropical Freshwater Zooplankton. *Identification, Ecology and Impact on Fisheries.* Backhuys Publishers, Leiden, The Netherlands. pp: 189-233.

In: Microcosms
Editor: Christopher C. Harris
ISBN: 978-1-62618-661-3

Chapter 7

FLOATING DISH MICROCOSMS: A VERSATILE TOOL TO STUDY THE DEVELOPMENT OF BIOFILM COMMUNITIES

***Wolf-Rainer Abraham*[1,*] *and Renata Pereira*[1,2]**
[1]Helmholtz Center for Infection Research,
Chemical Microbiology, Braunschweig, Germany
[2]Piracicaba Dental School - University of Campinas (UNICAMP), SP, Brazil

ABSTRACT

Bacteria in the environment are organized in communities and show often high degrees of worksharing and cooperation. They are resilient against noxious agents and their functions are redundant to compensate changing environmental conditions. Our understanding of these functional communities, however, is limited, especially because only a minority of strains present in the environment can currently be isolated. Microcosms are powerful tools for the study of microbial communities without the need for prior isolation of strains. We became especially interested in microbial communities growing on interfaces forming so called biofilms. To study functional communities organized in biofilms we developed floating dish microcosms. Here soil or sediment samples from the environment are placed in a pot, covered with sterile water and a dish carrying different materials underneath is floated on the surface. Bacteria now migrate through the water column to these substrata and form complex biofilms. We used these biofilms to i) isolate degraders of pollutants used as substrata (polychlorinated biphenyls, hexachlorocyclohexanes, tributyltinoxide), ii) monitor biofilm development over time observing biofilm structure and composition and we identified the succession of degraders of different congeners, iii) compare biofilm communities from different sites and found similar activities but different species composition from different sites and iv) manipulate biofilm composition by using different substrata or adding compounds known

* Corresponding author: Wolf-Rainer Abraham. Helmholtz Center for Infection Research, Chemical Microbiology, Inhoffenstrasse 7, 38124 Braunschweig, Germany. Phone: +49-531-6181 4300, Fax: +49-531-6181 4699, E-Mail: wab@gbf.de.

to control biofilm formation. The floating dish microcosms can easily be adapted to specific research requirements making the whole approach very flexible. We optimized the protocol so that the microcosms can be used to grow biofilm communities which can be characterized by confocal laser microscopy, phylogenetic fingerprinting of the community composition and chemical analysis of pollutant degradation. The microcosms can also be adapted for the screening of natural compounds or novel materials for their ability to shape biofilm composition, structure and activity. In this chapter we present different floating dish microcosms and their scientific results. We also discuss the potential but also the limitations these microcosms have.

INTRODUCTION

One of the fundamental characteristics of the microbial world is its inherent metabolic diversity and its adaptability to environmental stresses. The ubiquitous distribution of bacteria coincides with the observation that microorganisms, and their natural communities, are of fundamental importance for the global cycling of organic matter [1]. The metabolic diversity is reflected in the species diversity and, particularly, in the large evolutionary distances observed between the main phylogenetic lineages of microorganisms [2].

Moreover, one should recognise that the vast majority of microorganisms present in environmental samples currently cannot be cultivated in the laboratory. This limitation prevents the characterization of the taxa by ordinary physiological tests leaving us with almost no clue for the requirements of several phyla for which no isolates are available. An important question concerning the culturable microorganisms arises as to their actual frequencies and significance in the ecosystem [3]. The problem is further complicated due to the fact that in nature microorganisms live in complex communities whose compositions vary depending on the environmental conditions. An extraordinary physiological flexibility, achieved during, at least, 3.5 billion years of evolution and adaptation [4, 5], allows microorganisms not only to colonize all habitats which are inhabited by eucaryotes but also to exploit a wide spectrum of extreme biotopes (e. g., hot springs, salt and soda lakes, highly-polluted waste waters, etc.) which are inhospitable to higher organisms.

To survive under such hostile conditions like -70° or +113°C, pH 0.7 or 13, saturated salt solutions, high radiation or water insoluble carbon sources most microorganisms do not live as individual cells but in communities organized as pseudo-multicellular organisms usually attached to surfaces and interfaces [6].

These aggregates, here summarized under the term biofilms, offer protection against e. g. grazing, extreme pH [7], heavy metals, osmotic stress or noxious agents like oxidants [8, 9], antibiotics [10] or antibodies. Some other advantages of this life style are increased tolerance against environmental changes, the ability to exchange substrates and genetic information, formation of steep gradients within the community and release of enzymes and toxins into the matrix formed by the extracellular polymer substances (EPS) [11] which serve here as minireactor. The mineralization of many compounds and polymers cannot be achieved by a single strain; instead microbial consortia are cooperating in close proximity. To study these communities and consortia is of fundamental importance for the understanding and manipulation of biofilms and their microbial processes.

Controlling biofilms is one of the fundamental problems of many processes. On the one hand biofilms are important in cleaning rivers and lakes or in the treatment of our waste water. Here the optimization of the microbial communities for maximal performance is in the the center of many efforts. On the other hand biofilms can be detrimental degrading buildings or machines by several forms of corrosion. Of growing concerns are biofilms in medicine, e. g. on implants or catheters. They are difficult to treat with antibiotics because of a much higher resistance of pathogens inhabiting the biofilms [12].

Often the replacement of the implant is the only solution against these biofilm infections. In general the aim is therefore a manipulation of biofilms preventing their formation where they are dangerous but shaping their formation and community composition in applications where their metabolic activity is highly welcome. To achieve this goal we have to understand the interactions of microbial communities organized in biofilms.

Flexible Design of the Floating Dish Microcosms

Under the aspect that most bacteria can currently not be isolated from the environment and that microbial communities are prevailing in nature and forming highly interacting consortia we developed an approach for growing biofilm communities from environmental samples without any prior isolation of strains. We used here the concept of microcosms which are designed to maintain and to manipulate environmental samples in the laboratory without any isolation of their members. The aim here was that a simple but flexible system should be developed which can be miniaturized for parallel experiments.

To grow biofilm communities from soil or sediment samples the sample was placed in a sterile reservoir and covered with 3-5 cm of sterile tap water. After letting the sediment settle for one day a sterile dish made of Permanox® was placed at the surface of the water (Figure 1). Permanox® is a solvent-resistant polymer which is inert against many organics, e. g. polychlorinated biphenyls, tributyltin or hexachlorocyclohexane, although solvents such as xylene or toluene cause Permanox plastic slides to warp.

Because its autofluorescence at commonly used wavelengths is very low, Permanox® products are recommended for fluorescent applications. The microcosm with the floating dishes is kept at a constant temperature without agitation.

To follow the development of biofilms over time new dishes can be added at given time intervals, and all are harvested at the end of the experiment. This results in increasing incubation times giving biofilms of different states of maturation. The grown biofilm communities can then be analysed for i) biofilm architecture using laser scanning confocal microscopy (LSCM), ii) microbial community composition using various sequencing and fingerprinting techniques and iii) for chemical composition applying appropriate methods of organic analytics. This floating dish microcosm concept was used to grow biofilm communities from environmental samples. When the floating dish was loaded with water insoluble substrates biofilm communities can be grown on these substrates. This can be used to monitor the degradation of these compounds but also to enrich degraders of these substances from the biofilm communities. Finally, the composition of the developing biofilm communities can be modulated by exposing them to different substrata. In the following sections examples for these approaches will be given.

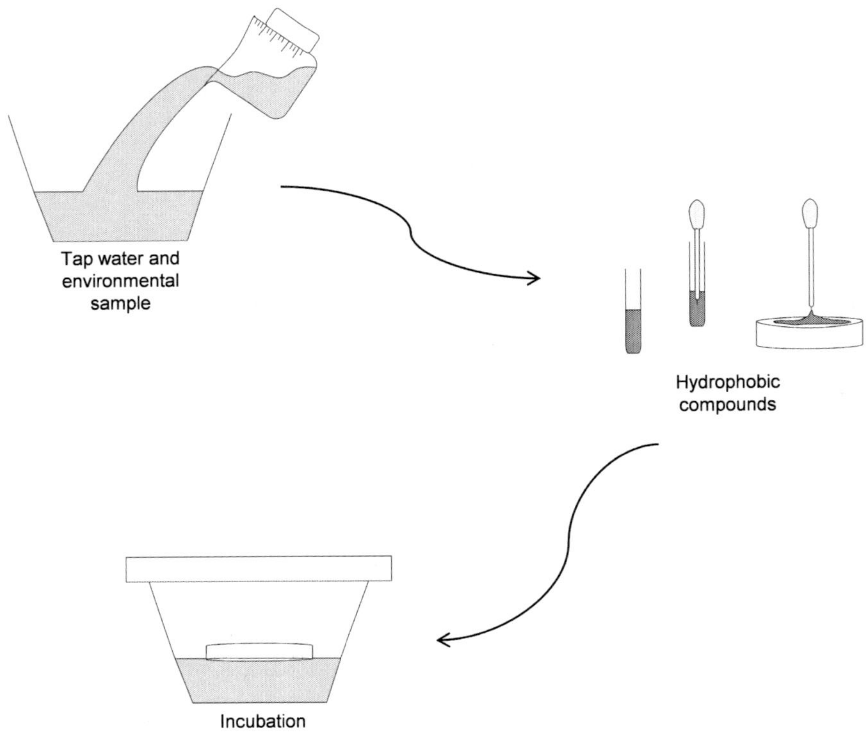

Figure 1. General scheme of floating dish microcosms. Tap water in a container is inoculated with the environmental sample. After waiting for 4-8 hours for the sedimentation of larger grains the floating dish is set on the water surface and the container is closed. The microcosm is incubated at a given temperature in still culture. After the experiment the dish with the biofilm community is removed and analysed.

ENRICHING BACTERIA FROM THE ENVIRONMENT

We used floating dish microcosms to study biofilm communities from a site heavily polluted with polychlorinated biphenyls (PCB). PCB comprises a class of substances consisting of 209 different compounds, called congeners, differing by the degree of chlorination and the position of the chlorine atoms at the aromatic rings. PCB has a number of remarkable physical properties which can be amplified by mixing different PCB congeners. The excellent physical properties of PCB and its extreme chemical stability were the reasons for its broad industrial application. However, the same properties make PCB a major environmental pollutant as well and allow its enrichment in food chains. All congeners are poorly soluble in water but reasonably good soluble in organic solvents and in fat allowing the enrichment of PCB in food chains. PCB was produced for decades for a wide variety of uses but banned when it became obvious that it is also a cancerogen, immune modulator and hormone signalling disruptor. Nevertheless, it is still present in the environment and an important health risk [13].

We were interested how bacteria deal with these hydrophobic substrates in soil. We added soil samples from a site near Wittenberg, Germany, polluted with PCB into a vessel, covered it with tape water and placed a Permanox dish on the water surface.

The idea of this approach was that the very complex bacteria community found in the soil samples [14, 15] will be reduced to those bacteria which are able to interact and form a biofilm on the Permanox substratum. When we examined the Permanox surface by microscopy we were surprised to see no bacterial cells but small aggregates of clay particles. Doing the same experiments with soil autoclaved before being used in the experiment prevented the formation of these clay aggregates. This pointed to an involvement of bacteria in this process. These bacteria were finally found in ultrathin sectioned clay aggregates revealing 1-4 bacteria cells attached to the Permanox surface and completely covered by clay minerals. The function of these clay aggregates, dubbed "clay-hutches", became apparent when soil samples from the same site but lacking any detectable contamination with PCB were used in the experiment. From these clean soil samples no clay aggregates could be detected [16]. Using culture-independent methods based on the diversity of 16S rRNA gene sequences the main bacteria species responsible for the formation of the clay aggregates were identified. This was achieved by T-RFLP and comparison of the fragment size with predicted chain length from the 16S rRNA gene database. The diversity of the soil samples was similar independent from their PCB pollution. In contrast, the evenness decreased with increasing PCB concentration in the soil samples, indicating that a few bacterial groups had adapted to PCB and became dominant within the bacterial community. These were mainly members of the genus *Burkholderia*, and some *Pseudomonas* spp. In the biofilm derived from the most heavily polluted soil sample, *Burkholderia* spp. became dominant. Also members of the *Acidobacterium-Holophaga* phylum were found probably due to the low pH of 4.5 of the soil samples but became less prominent in PCB polluted soils. In contrast, relatives of the genere *Nevskia* and *Delftia* contributed constantly to all biofilms [17]. From these results it has been hypothesized that bacteria attach the clay particles which are heavily loaded with PCB to their cells to get access to this water insoluble substrate. For many years it was completely unclear how bacteria access the hydrophic PCB and transport it into their cells for mineralization. One answer to this question seems to be these clay-hutches where bacteria catch the PCB-soaked clay particles when they are moved around by running water, e. g. after a rain fall. The observed clay aggregates are one example of composite biofilms and floating dish microcosms are a versatile way to study them under different growth conditions in the laboratory [18].

DEGRADATION OF POLLUTANTS

The approach to grow biofilm communities from polluted soil and sediment samples can be modified by putting the pollutant on the water-facing side of the Permanox dish and grow biofilm communities directly on the pollutant. This works for all hydrophobic pollutants which form then both a substratum and a substrate for the developing biofilm communities. We have done this for PCB and used several mixtures of PCB congeners to obtain biofilm communities. The aim at this stage was to use this approach for the enrichment of metabolically active bacteria. When the biofilm developed on the PCB droplet it was plated on agar dishes to isolate PCB-degrading bacteria. The metabolic activity of bacteria isolated from the biofilm communities were generally higher than that of strains isolated directly from the soil used as inoculum for the biofilm. Thus, the biofilm communities can be regarded as an additional enrichment step for metabolic active bacteria. These results have been confirmed by using polycyclic aromatic hydrocarbons (PAH) instead of PCB as substrate [19]. The approach is limited by the solubility of the substrate in water. However, even hexachloro-cyclohexanes (HCH) gave good results although the crystals on the Permanox slides dissolved in the medium after about 3 days [20].

DYNAMIC OF BIOFILM COMMUNITIES

Biofilms are not static structures but develop over time and adapt rapidly to changing environmental conditions. Floating dish microcosms can also be used to study the dynamic of the developing biofilm communities. We have shown this for biofilms growing on PCB. Permanox dishes exposed to soil samples polluted with PCB were harvested at different time points in order to assess the developing biofilm communities.

The individual biofilm communities were analysed for their species composition using single strand conformational polymorphism (SSCP) profiling and sequencing of the main bands of the gels. The architecture of the biofilms was determined using confocal Laser scanning microscopy (CLSM) [21]. To highlight differences in the biofilms a variety of staining protocols [22] were applied revealing an astonishing diversity of microstructures within the biofilm communities (Figure 2).

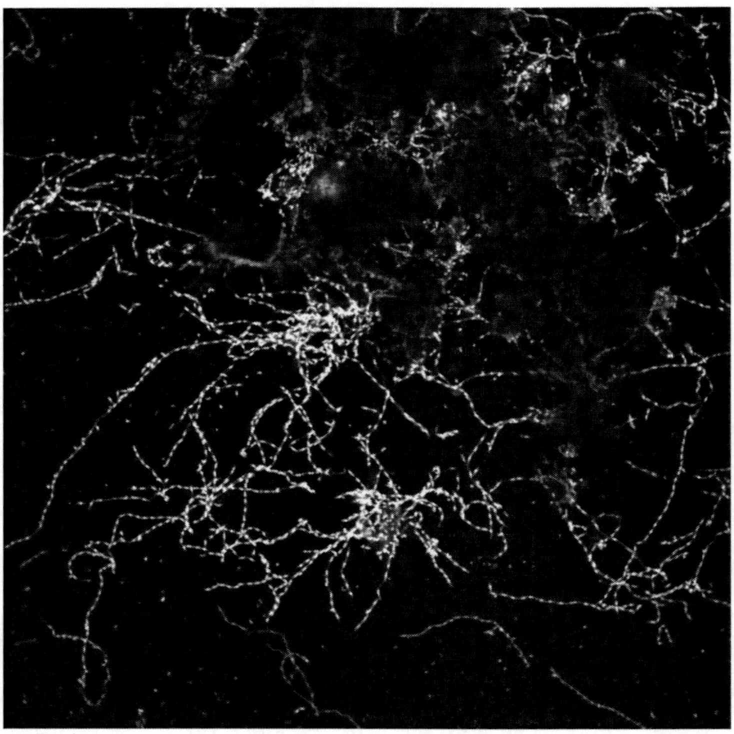

Foto: Maximiliano Gutierrez.

Figure 2. Biofilm community grown after 2 days on hexachlorocyclohexane microcrystals coated on a Permanox® dish. The microcosm was inoculated with river sediments. Staining of the sample was achieved by Live/Dead® stain (Molecular Probes).

Surprisingly, a very dynamic biofilm was found which formed first not at the PCB droplet itself but around the droplet. This surrounding biofilm then moves to the PCB droplet and covers it completely [23].

The final stage of biofilm development is characterized by the invasion of bacteria into the PCB oil forming holes and entering anaerobic growth conditions. This facilitates the degradation of highly chlorinated PCB congeners by reductive dechlorination which requires anaerobic conditions. Instead of comparing biofilms from the same soil sample the assessment of the developing biofilm communities can also be used to compare the metabolic potential of different soil samples. Doing this with soil samples of no history of PCB pollution showed the widespread capacity to metabolize polychlorinated biphenyls by diverse microbial communities in these soils with no significant exposure to PCB contamination [24]. It can be speculated that these metabolic capacities offer a selective advantage for the cells if PCB is eventually spilled to the soil.

We have confirmed this assumption by comparing the metabolic activities of biofilm communities obtained from soil samples of various degree of PCB pollution. An impressive adaptation of PCB-degraders to highly polluted soils could be shown revealing a much faster and more complete degradation of PCB by microbial communities from highly contaminated soil [25].

For all biofilms complex microbial communities were observed. But who in these communities is actually doing the degradation of PCB? This question is usually answered by enriching bacteria on the substrate. Doing this with the given soil samples always revealed strains of the genus *Burkholderia* as the main degrader.

However, it has been shown for several soil samples that the vast majority of bacteria can currently not be cultivated simply because we do not know their specific growth conditions. Attempts to isolate the main degraders therefore may miss this uncultured majority and to avoid such a bias, culture-independent methods are required.

One way to assess metabolic activities of microbial communities is the application of substrates labelled with stable isotopes. Such substrates do not bring additional risks to the environment because stable isotopes are not radioactive. The substrates are degraded and the stable isotope is integrated only into the biomass of the degrader. Enrichment of the biomass by the added stable isotope demonstrates degradation of the substrate by someone in the community. If taxon-specific compounds, so called biomarkers, are analyzed, the responsible degrader can be narrowed down. For ^{13}C-labelled substrates these biomarkers may be specific fatty acids found only in some bacterial genera. Such an analysis is usually done by a separation of the methyl esters of these fatty acids (FAMEs) in a gas chromatograph where in an interface the separated FAMEs are oxidized by a catalyst to carbon dioxide which is finally analysed for it isotopic content in ^{13}C in an isotope ratio mass spectrometer (IRMS) [26]

The sensitivity and accuracy of such an instrument is nowadays good enough to see even isotope discriminations caused by individual enzymes [27]. We applied such an approach to a PCB degrading microbial community fed with a ^{13}C-labelled PCB-congener. The pattern of labelled fatty acids matched with the main fatty acids of *Burkholderia* species identifying strains of this bacterial genus as the main degraders of PCB [28].

BIOFILM MANIPULATIONS IN FLOATING DISH MICROCOSMS

All biofilms discussed so far are beneficial because they can be applied to clean the polluted environment. However, not all biofilms are always welcomed. Especially biofilms growing on or in the human body are often problematic and the source of chronic infections. Biofilm infections on implants can cause severe complications for the patients and high health costs. The occurrence of these infections is dependent on the implant but also on the duration of the required surgery.

Considering the strongly enhanced resistance of biofilm communities against antibiotics, biofilm infections are usually difficult to treat. To overcome this limitation many attempts were made to inhibit biofilm formation in general. For medical implants the obvious target for such an approach would be the search for antimicrobial surfaces. Here a wealth of results have been published and many tailored surfaces have been patented which hinder or even prevent the colonization by specific pathogens.

Most of these materials, however, have usually two main problems: On the one hand they prevent not only colonization by bacteria but also the integration of the implant into tissue and on the other hand these materials are usually very specific in their antimicrobial activities preventing colonization by certain pathogens but being completely inert for other microorganisms. The situation is especially problematic for oral implants which are continuously exposed to a multitude of bacteria in the oral cavity [29]. Here often a screening of several materials is necessary as models for biofilm prevention are still rather non-specific severely limiting a rational design of antimicrobial coatings.

We adapted the floating dish microcosms to allow the screening of several substrata using the same clinical inoculum. We wanted to use swabs from dental biofilms from volunteers but were confronted with the problem that each swab is unique which can cause problems in the comparison of different substrata. We solved this problem by mounting different substrata at the floating dish and exposing them together under identical conditions to the same microbial community (Figure 3). We used titanium plates coated with various copolymers in order to identify those copolymers which prevented best the formation of biofilms on the substratum [30].

We observed for some co-polymers a significant reduction of biofilm formation while others had no or only little effect on the cell densities of the biofilms. An interesting phenomenon is a certain selection of the co-polymers for some bacteria. These bacteria are suppressed in the biofilm communities by only some co-polymers, but also an enhancement of other strains could be found (Figure 4). This opens the possibility of modulation of biofilm communities by specific coatings of implants with the aim to suppress specifically pathogens in the biofilm communities.

Furthermore, by exposing several substrata and harvesting them at different time points a kinetic of biofilm formation could be determined.

This includes both the cell density, the developing architecture of the biofilm but also the succession of bacteria in the developing biofilm communities.

Although the swaps always introduced unique microbial communities into the microcosms the performance of the different copolymer-covered substrata was the same, i. e. a copolymer with low biofilm formation always gave weak biofilms regardless of the inoculum.

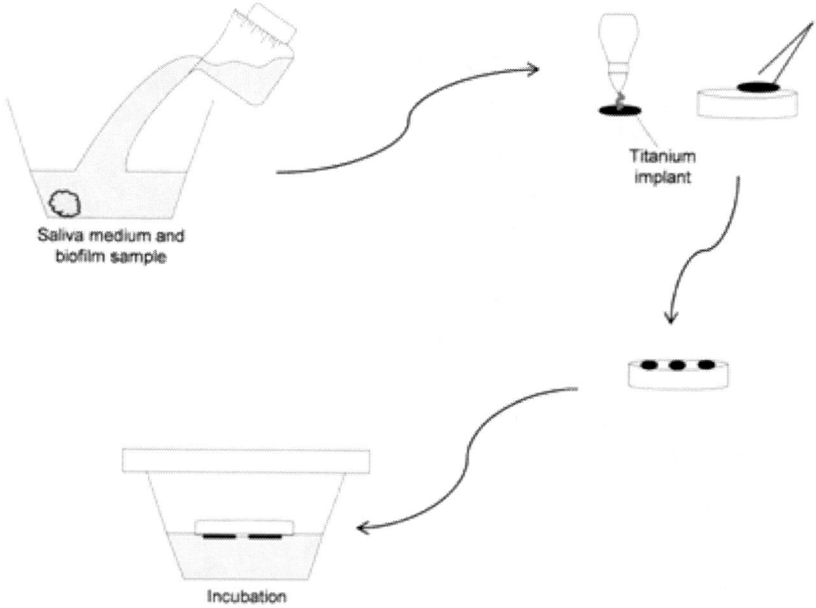

Figure 3. Scheme of the screening of antimicrobial coatings with the floating dish microcosm. The container is filled with artificial saliva medium and inoculated with swaps from teeth of a volunteer. Titanium plates with different antimicrobial coating are attached to a Permanox® dish which floats on the surface of the medium. The container is closed and incubated at 37°C without shaking. Exposure of the titanium plates to the same inoculum under identical conditions allows direct comparison of biofilm communities grown on the different coatings.

CONCLUSION

In nature bacteria rarely occur as pure strains but prefer to live in microbial communities. Many of these communities are attached to surfaces forming self-organizing biofilms. These biofilm communities have a multitude of interactions, e. g. forming steep gradients of nutrients, pH, oxygen, salt, harbouring microcolonies [31] and pathogens and communicating between the individual bacterial cells and with the eukaryotic host by small molecules [32]. Any approach to modulate biofilm communities has to take these complex interactions into account which cannot be simulated by single-strain experiments. It has widely been noticed that many antimicrobial coatings showing impressive results against certain pathogens failed completely in animal models because biofilm community interactions have not been considered. To overcome this limitation simply mixing different bacteria strains to obtain biofilm community conditions does not work because a subtle homeostasis between individual community members controls microbial communities. Only sophisticated protocols allow the generation of stable microbial communities from pure strains but even if such a community is successfully generated it lacks many bacteria present in the environment but being resistant against current isolation procedures. These currently nonculturable bacteria still form the vast majority of all species present in the environment exceeding 99% of all cells in many habitats [33]. To generate biofilm communities and include also nonculturable bacteria a microcosm approach turned out to be a promising one.

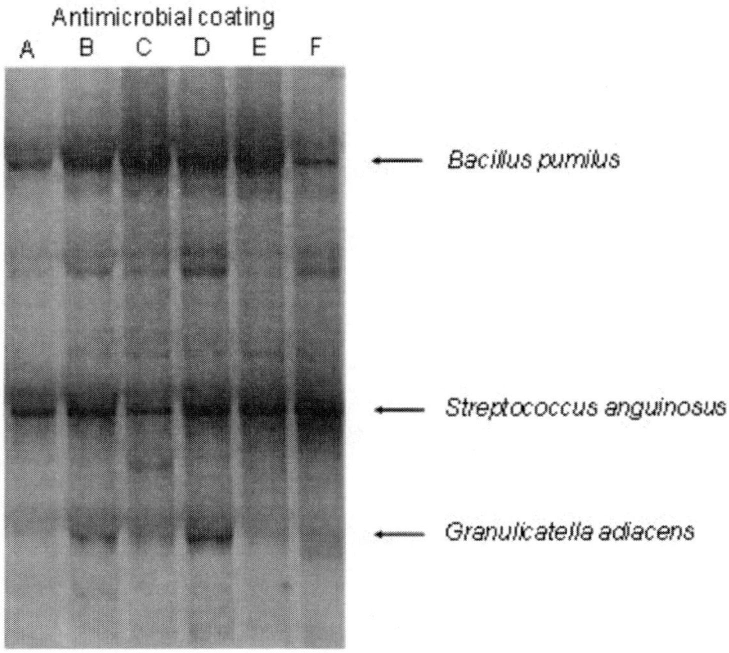

Figure 4. Biofilm communities grown on antimicrobial coatings A - F. SSCP gel based on amplicons of the 16S rRNA genes. Equal amount of DNA for all lanes has been applied to the gel, therefore, differences in band intensities stand for differences in species cell abundances. Each band represents a different species; bands at the same heights are the same species. The main bands have been extracted, sequenced and identified via comparison with databases. It is obvious that similar biofilm communities developed on the different coatings, however, some differences in species abundances between the different substrata can be found.

The general goal to study biofilm communities under different conditions requires a flexible design of the microcosms. This was achieved by inoculating tap water with the environmental sample of interest and exposing suitable substrata for biofilm formation to the medium. Floating the substrata at the surface of the water gave well reproducible biofilms and allowed easy removal of the substrata for analysis.

Using Permanox slides as substrata enables a direct investigation of the attached biofilms by CLSM and selective staining highlighting characteristics of the extracellular matrix surrounding the bacterial cells. Extracting total DNA from the biofilms and applying fingerprinting methods reveals community composition.

These basic features of floating dish microcosms can be adapted to study special aspects of biofilms communities. Exposure of organic hydrophobic substrates to the inoculum leads to biofilm communities consuming these substrates and labelling these substrates with stable isotopes helps to identify the main degraders of these substrates.

These studies open new aspects in bioremediations of polluted sites using endogenous biofilm communities.

Instead of substrates antimicrobial substrata can be exposed to the medium and the developing biofilms can be quantified by cell numbers and by the selection of bacteria able to colonize these substrata.

One should, however, also be aware that floating dish microcosms have their limitations. First of all they are still models mirroring conditions in the environment in an imperfect way.

Thus, results from these microcosms always need confirmation by *in vivo* or *in situ* experiments. Furthermore, the current approach requires water-insoluble substrates for biofilm formation. Attempts to include water soluble substrates by absorption on a carrier or by diffusion-release did not give satisfactory results. To deal not with forming biofilms but with already established ones is more realistic for the treatment of clinical biofilms. Addition of quorum-quenching or other biofilm-dissolving compounds to biofilms grown in floating dish microcosms would be an attractive approach to control these biofilms. The limitation here is that biofilms grown from environmental, here clinical, samples are variable giving often unique microbial communities. This limitation has to be overcome in order to screen potential anti-biofilm drugs.

Despite these limitations floating dish microcosms open a broad spectrum of biofilm studies in a wide variety of habitats and applications. The set-up of these microcosms is easy and highly flexible which allows studies of many aspects of biofilms closing the gap between results from pure culture experiments and *in vivo* or *in situ* studies.

ACKNOWLEDGMENTS

Parts of this study were done within the Collaborative Research Centre SFB 599 supported by a grant of the German Research Council (DFG). R. P. acknowledges the support by a stipend of CNPq (Brazil).

REFERENCES

[1] Cunliffe, M., R. C. Upstill-Goddard and J. C. Murrell. 2011. Microbiology of aquatic surface microlayers. *FEMS Microbiol. Rev.* 35: 233-246.

[2] Woese, C. R. 1987. Bacterial evolution. *Microbiol. Rev.* 51:221-271.

[3] Amann, R. I., W. Ludwig and K. H. Schleifer. 1995. Phylogenetic identification and in situ detection of individual microbial cells without cultivation. *Microbiol. Rev.* 59:143-169.

[4] Doolittle, R. F., D. F. Feng, S. Tsang, G. Cho, and E. Little. 1996. Determining divergence times of the major kingdoms of living organisms with a protein clock. *Science* 271:470-477.

[5] Pace, N. R. 1997. A molecular view of microbial diversity and the biosphere. *Science* 276:734-740.

[6] Tice, M. M. and D. R. Lowe. 2004. Photosynthetic microbial mats in the 3,416-Myr-old ocean. *Nature* 431:549-552.

[7] Li, Y. H., M. N. Hanna, G. Svensater, R. P. Ellen, and D. G. Cvitkovitch. 2001. Cell density modulates acid adaptation in *Streptococcus mutans*: implications for survival in biofilms. *J. Bacteriol.* 183:6875-6884.

[8] Hassett, D. J., J. F. Ma, J. G. Elkins, T. R. McDermott, U. A. Ochsner, S. E. West, C. T. Huang, J. Fredericks, S. Burnett, P. S. Stewart, G. McFeters, L. Passador, and B. H. Iglewski. (1999) Quorum sensing in *Pseudomonas aeruginosa* controls expression of

catalase and superoxide dismutase genes and mediates biofilm susceptibility to hydrogen peroxide. *Mol. Microbiol.* 34:1082-1093.

[9] Walker, J. T., D. J. Bradshaw, M. R. Fulford, and P. D. Marsh. 2003. Microbiological evaluation of a range of disinfectant products to control mixed-species biofilm contamination in a laboratory model of a dental unit water system. *Appl. Environ. Microbiol.* 69:3327-3332.

[10] Stewart, P. S. and J. W. Costerton. 2001. Antibiotic resistance of bacteria in biofilms. *Lancet* 358:135-138.

[11] Davey, M. E. and G. A. O'Toole. 2000. Microbial biofilms: from ecology to molecular genetics. *Microbiol. Mol. Biol. Rev.* 64:847-867.

[12] Macedo, A. J. and W.-R. Abraham. 2009. Can infectious biofilms be controlled by blocking bacterial communication? *Med. Chem.* 5:517-528.

[13] Kamrin, M. A. and R. K. Ringer. 1994. PCB Residues in mammals: A review. *Toxicol. Environ. Chem.* 41:63-84.

[14] Nogales, B., E. R. Moore, E. Llobet-Brossa, R. Rossello-Mora, R. Amann, and K. N. Timmis. 2001. Combined use of 16S ribosomal DNA and 16S rRNA to study the bacterial community of polychlorinated biphenyl-polluted soil. *Appl. Environ. Microbiol.* 67:1874-1884.

[15] Nogales, B., E. R. Moore, W.-R. Abraham, and K. N. Timmis. 1999. Identification of the metabolically active members of a bacterial community in a polychlorinated biphenyl-polluted moorland soil. *Environ. Microbiol.* 1:199-212.

[16] Lünsdorf, H., R. W. Erb, W.-R. Abraham, and K. N. Timmis. 2000. "Clay hutches": a novel interaction between bacteria and clay minerals. *Environ. Microbiol.* 2:161-168.

[17] Abraham, W.-R., H. Lünsdorf, C. Strömpl, B. Nogales, E. R. B. Moore, and K. N. Timmis. 2003. Microbial communities in composite biofilms participating in the degradation of PCB. *Water, Air, Soil Pollution: Focus: Bioremediation* 3:57-64.

[18] Lünsdorf, H., C. Strömpl, A. M. Osborn, A. Bennasar, E. R. B. Moore, W.-R. Abraham, and K. N. Timmis. 2001. Approach to analyse interactions of microorganisms, hydrophobic substrates, and soil colloids leading to formation of composite biofilms, and to study initial events in microbiogeological processes. *Methods Enzymol.* 336:317-331.

[19] Heck, M. G. 2011. Functional biodiversity of *Pseudomonas* species in biofilm communities degrading polycyclic aromatic hydrocarbons. *Doctoral thesis*, Technical University Brauschweig, Germany.

[20] Gebreil, A. S. 2011. Diversity of microbial communities in biofilms growing on hexachlorohexane and related substrates. *Doctoral thesis*, Technical University Braunschweig, Germany.

[21] Macedo, A. J. and W.-R. Abraham. 2009. Microcosms for biofilm analysis on hydrophobic substrates – A multiple approach to study biodiversity, metabolic activity and biofilm structure and dynamic. In: *Handbook of hydrocarbon microbiology: Microbial interactions with hydrocarbons, oils, fats and related hydrophobic substrates and products.* Ed. K. N. Timmis, Springer, 3544-3551.

[22] Neu, T. R., U. Kuhlicke and J. R. Lawrence. 2002. Assessment of fluorochromes for two-photon laser scanning microscopy of biofilms. *Appl. Environ. Microbiol.* 68:901-909.

[23] Macedo, A. J., U. Kuhlicke, T. Neu, K. N. Timmis, and W.-R. Abraham. 2005. Three stages of a biofilm community developing at the liquid-liquid interface between polychlorinated biphenyls and water. *Appl. Environ. Microbiol.* 71:7301-7309.

[24] Macedo, A. J., K. N. Timmis and W.-R. Abraham. 2007. Widespread capacity to metabolize polychlorinated biphenyls by diverse microbial communities in soils with no significant exposure to PCB contamination. *Environ. Microbiol.* 9:1890-1897.

[25] Macedo, A. J., U. Kuhlicke, T. Neu, and W.-R. Abraham. 2006. Adaptation of microbial communities in polychlorinated biphenyls contaminated soil leading to the degradation of higher chlorinated congeners in biofilms. *Biofilms* 3:37-46.

[26] Abraham, W.-R., C. Hesse and O. Pelz. 1998. Ratios of carbon isotopes in microbial lipids as indicator of substrate usage. *Appl. Environ. Microbiol.* 64:4202-4209.

[27] Abraham, W.-R. and C. Hesse. 2003. Isotope fractionations in the biosynthesis of cell compounds by different fungi: A basis for environmental carbon flux studies. *FEMS Microb. Ecol.* 46:121-128.

[28] Tillmann, S., C. Strömpl, K. N. Timmis, and W.-R. Abraham. 2005. Stable isotope probing reveals the dominant role of *Burkholderia* sp. in aerobic degradation of PCBs. *FEMS Microb. Ecol.* 52:207-217.

[29] Esposito, M., M. G. Grusovin and H. V. Worthington. 2012. Interventions for replacing missing teeth: treatment of peri-implantitis. Cochrane Database Syst. Rev. 1:CD004970. doi: 10.1002/14651858.CD 004970.pub5.

[30] Pfaffenroth, C., A. Winkel, W. Dempwolf, L. J. Gamble, D. G. Castner, M. Stiesch, and H. Menzel. 2011. Self-assembled antimicrobial and biocompatible copolymer films on titanium. *Macromol. Biosci.* 11:1515-1525.

[31] Lünsdorf, H., I. Brümmer, K. N. Timmis, and I. Wagner-Döbler. 1997. Metal selectivity of *in situ* microcolonies in biofilms of the Elbe river. *J. Bacteriol.* 179:31-40.

[32] Estrela, A. B., M. G. Heck and W.-R. Abraham. 2009. Novel approaches to control biofilm infections. *Curr. Med. Chem.* 16:1512-1530.

[33] Jones, J. G. 1977. The effect of environmental factors on estimated viable and total populations of planktonic bacteria in lakes and experimental enclosures. *Freshwater Biol.* 7:67-91

In: Microcosms
Editor: Christopher C. Harris
ISBN: 978-1-62618-661-3

Chapter 8

Cohesion and Sustainability Inside Microcosms: The Role of Semiosis

Maria Isabel Aldinhas Ferreira*

Centre of Philosophy of the University of Lisbon.
Language, Mind and Cognition Group
Faculdade de Letras. University of Lisbon, Lisbon, Portugal

Abstract

The dialectic relationship that characterises the semiotic process inherent to any form of cognition is responsible for the cohesion and sustainability of the microcosms defined by living entities in their interaction with the surrounding world.

The coupling that arises as a result of the plasticity of both living organism and its environmental bubble produces an autonomous and strictly bounded unit that has nevertheless been shaped extensively in the course of its evolutionary and developmental history.

Semiosis is consequently an embodied, embedded and always situated phenomenon. This means it involves a living entity endowed with a particular physical architecture interacting with the specific world it is immersed in, producing a dynamics that defines events taking place in a particular microcosm and perceived by a human observer as anchored in space/time.

Uexkull (1909) introduced the fundamental concepts of Umwelt and Innenwelt stressing their strict interdependent nature. The present paper stems from these fundamental concepts but it goes further by equating them with the semiosis inherent to any form of cognition. Plus, by proposing the mathematical modelling of a prototypical instance of individual cognition and by introducing the concept of Umwelt Overlap the paper attempts to identify the essential parameters of ecological sustainability.

Keywords: Cognition, Microcosm, Semiosis, Umwelt, Innenwelt, Cohesion, Umwelt Overlap, Ecological Sustainability

* E-mail: aldinhasferreira@gmail.com.

1. Semiosis: An Embodied, Embedded and Situated Phenomenon

A life form and its environment constitute a unit, a microcosm. The cohesion and sustainability of this microcosm is guaranteed by an essential semiotic process that is actualised in the privileged dialectic relationship established between the individual living entity and the environmental bubble this entity is embedded in.

This dialectic relationship that characterises the semiosis inherent to any form of cognition[1] is responsible for the cohesion and sustainability of the microcosms defined by living entities in their interaction with the surrounding world.

Every species has a typical evolved architecture, the phenotypic structure, which is the joint product of its genes and the environmental variations the organism had to face during its evolutionary and developmental history.

In what relates evolved systems, form seems to follow from function, as the existence of a particular physical structure is determined by the specific functional needs the organism has met along its evolutionary history. This functional level of explanation is essential to understand how natural selection designs organisms and how, in the course of evolutionary time, new design features were added or discarded from the species design.

Genetic "instructions" provide general constraints for neural development determining the different levels of neural organisation and the specificity of sensorial equipment that organisms belonging to different species display constraining this way the types and forms of interaction available and consequently the nature of the organism's particular responses to environment.

On this account Varela (1992) refers:

> "The nature of the environment [...] acquires a curious status: it is that which lends itself (es lehnt sich an...) to a surplus of significance. Like jazz improvisation, environment provides the "excuse" for the neural "music" from the perspective of the cognitive system involved".

The capacity to identify specific environmental features springs out naturally when the life form interacts with the external world, defining specific semiotic relations. That disposition seems to rest upon a significant degree of innate "knowledge"[2]- a "know how", which all organisms possess, and which is, again, the result of the experience of biological predecessors and a consequence of their adaptive efforts to adequately respond to

---

[1] Cognition is here understood as a qualitative property not only emergent in organisms, such as mammals, but rather the capability of every living or artificial system to cope with its environment by adequately interpreting and responding to environmental stimuli. This adequacy of response comprehends not only the system's capacity to retrieve from the surrounding environment all the information necessary to its maintenance. It also involves the capacity of intervening in that same environment creating the most favourable niches and the capacity of adapting itself through morphological or structural changes when the parameters that define its adequate environment change radically. Cf. Maturana (1978) and Maturana and Varela (1980) on this subject.
In what relates human cognition that capacity of response relates not only the quality of interaction with a physical environment but also the quality of the individual's participation in all the social and cultural spheres in which s/he evolves.

[2] In our opinion the term *knowledge* designates information that is conceptually and propositionally structured. Consequently, we refer to the "know how" that is apparently embodied in the distinct organic architectures as "knowledge".

environmental conditions and changes. Based on the recurring properties of previous encounters, these architectures embody vital information concerning the typical environmental patterns the living entity will have to face, guaranteeing this way the "know how" that guides all its actual interactions. This "know how" comprehends the capacity to identify and interpret specific environmental features triggering out the organism's adequate response from a "pack" that is already pre-established.[3]

This innate "know how" implicit in the different forms organisms interact with the external world is also present in the way the various subsystems inside those organisms are able to individuate, identify and assign meaning to certain features in their secluded contexts, guaranteeing the normal functioning of the vital systems that regulate the entire organic structure of the body. An example of what has just been said is the form how homeostatic systems function. In fact homeostatic systems require cell-to-cell communication via chemical messengers. The first step in the action of a chemical messenger is its binding to specific target-cell proteins known as receptors. These receptors are selective in what relates the type and number of chemical messengers. Although a chemical messenger (a hormone, a neurotransmitter, a paracrine/autocrine agent, or a plasma-membrane-bound messenger) may come into contact with many different cells, it influences only certain cells and not others as only certain cell types, frequently just one, possess the receptor required for the combination with a particular chemical messenger. The combination of chemical messenger and receptor initiates the events leading to the cell's ultimate responses.

In this sense we can agree with Merleau-Ponty (1962) in that meaning exists at a pre-reflective level of existence. In fact there seems to be a primary, pre-ontological assignment of meaning, an assignment that is prior to any experience whatsoever and that enables to respond adequately to some environmental cues and to ignore others. This pre-objective understanding that seems to be built into all life forms is responsible for their functioning and determines the way they interact with the surrounding environment.[4]

At the core of all life processes lies semiosis, the essential life phenomenon of identification and meaning assignment to fundamental environmental features by a living entity.

Cognition demands constant awareness, the capacity of identifying and assigning a value-meaning- to specific environmental features immediately reacting in an adequate way. The promptness of response that characterises active sensors is accompanied by another important feature: selectiveness. The system recognises and responds only to stimuli that are meaningful and consequently "visible", i.e., that correspond to the potential satisfaction of its specific needs.

The selectiveness that characterises all forms of cognition is structure determined and apparently requires no effort on the part of the system involved. This selectiveness takes place because, as we pointed out before, living systems are genetically equipped to react affirmatively, under certain conditions, to clusters of stimuli, selecting qualities out of the

---

[3] Emmeche (2000) points out that the endosemiotic sign links can be analysed as causal links between the function parts that regulate the activity of the entire organism.

[4] Merleau- Ponty (1968) stresses that in what concerns human organisms this pre-objective understanding enables us to accomplish familiar tasks without making a conscious breakdown of distances or measurements before taking action. In "The Visible and the Invisible" the author points out:

"We are able to make these habitual movements because the space in which normal imitation operates is not an objective space or a representative space based on an act of thought. It is already built into my bodily structure and its inseparable correlative".

environment that correspond to specific purposes. In response to those specific needs, some qualities become salient or visible while all the others, because they are irrelevant, constitute "noise" that is ignored.

As posited by Ferreira (2007), (2011), independently of the type of cognition or the level of semiotic complexity involved, meaning is always a value- a structured entity. This value is assigned by the cognitive agent to an individuated environmental feature or to a cluster of features that, because of the agent's physical nature, emerge in the environment as salient.

The formula below, Ferreira (2011:25), attempts to model in mathematical terms the constitution of specific meaningful worlds:

> " Be *X* the set of all possible cognitive agents endowed with a specific physical architecture (domain *X)* and be *Y* the set of all possible environmental features (codomain *Y)* then *the potential meaningful world for X* could be translated as $f(x) \subseteq Y$".

In the diagram, figure 1, the oval on the left represents the set of all cognitive agents endowed with a particular physical architecture $\{X\}$, while the oval on the right represents the set of all possible environmental features $\{Y\}$. *f* is a function from domain *X* to codomain *Y*; the small oval stands for the image of *f,* i.e.the set of all possible outputs obtained when the function is evaluated at each element of the subset. In other words the smaller oval represents the set of all possible meaningful features for *X* in the codomain *Y*.

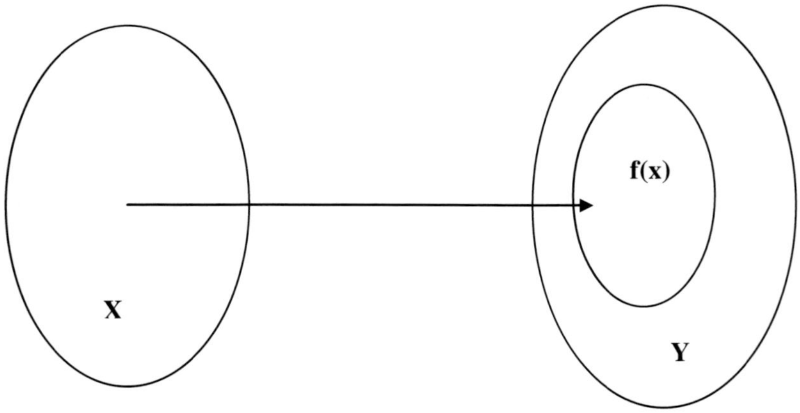

Figure 1. Meaning is the value assumed by f(x) in the codomain Y.

Uexküll (1926) believed that access to the different worlds of individual organisms could only be reached through the study of their specific organisation. This access demanded, according to him, that we submerge ourselves in the anatomical structure responsible for defining the way it interacts with the external world, and at the same time that we make as clear as possible the extent of the achievements that this structure brings about, in order to define the field of its existence and its activities.

The essential role played by the physical structure of organisms, namely by the sensorial apparatus, in the definition of a meaningful world, was stressed by Uexküll in these terms (1926: 126)

> "Every animal is a subject, which, in virtue of the structure peculiar to it, selects stimuli from the general influences of the outer world, and to these it responds in a certain way.[...] The sum of stimuli affecting an animal forms a world in itself."

Acknowledging that the physical structure of a particular life form is responsible for the definition of its specific meaningful world- Umwelt-, Uexküll highlights the specificity of each life form's experience and view of the world . In the introduction to "A Stroll in the World of Animals and Humans", this author refers (1934:5):

> "This little monograph does not claim to point to a new science. Perhaps it should be called a stroll into unfamiliar worlds; worlds strange to us but known to other creatures, manifold and varied as the animals themselves. The best time to set out on such an adventure is on a sunny day. The place, a flower- strewn meadow, humming with insects fluttering with butterflies. Here we may glimpse the worlds of the lowly dwellers of the meadow. To do so, we must first blow, in fancy, a soap bubble around each creature to represent its own world, filled with the perceptions which it alone knows. When we ourselves then step into one of these bubbles, the familiar meadow is transformed. Many of its colourful features disappear, others no longer belong together but appear in new relationships. A new world comes into being. Through the bubble we see the world of the burrowing worm, of the butterfly, or of the field mouse; the world as it appears to the animals themselves, not as it appears to us"

By inviting us into an imaginary stroll, Uexküll wants to stress how a particular physical environment, defined in human terms as a "flower-strewn meadow", is individually appropriated and differently constructed by the plethora of organisms that coexist there[5].

The metaphor of the soap bubble is particularly successful. The virtual sphere, the figurative perimeter, traced according to the nature of the organism's interactions with its surrounding world, visualises not only the scope of these interactions but it defines in the general environment the organism's Umwelt, i.e. its meaningful world.

The transparency of the bubble allows us to be conscious of that sphere still retaining sight of the overall scenario in all its vivid richness and complexity. The transparency of the bubble shows how the organism's world is in fact nested into a general environment where other "soap bubbles"- individual Umwelten- coexist.

One might think that a glass bubble could be an identically well-suited metaphor – enclosure plus transparency would also be guaranteed. But it was not by chance that Uexküll chose a soap and not a glass bubble as metaphor. Though transparent the glass bubble would exhibit rigid, inflexible boundaries, totally inadequate to express the plasticity demanded to the organism's Umwelt.

Instead, the boundaries defined by the soap bubble are flexible allowing its slight remoulding when parameters such as internal or external pressure, surface tension and temperature change. The plasticity of the soap bubble is perfect to express the dynamics of the coupling organism/environment or its recoupling when essential parameters vary and adaptation or evolution occur. On the other hand, the fact that soap bubbles are able to remain intact when touching each others' surfaces or eventually merging, defining pairs or small clusters, visualises the way distinct microcosms-different Umwelten-do not independently coexist but in fact crisscross or even overlap.

---

[5] The Gibsonian tradition uses identical imagery to introduce the concept of "affordance".

The type and quality of the interactions and consequently the definition of its constructed world depend, as it has been pointed out, on the physical nature of the living system. It is also the specific nature of different living forms that determines their differentiated needs in existential terms. Consequently, the system only identifies and responds to environmental features that potentially satisfy those specific needs or those that appear as a menace to their satisfaction or to existence itself.

Uexkull's (1909) distinguishes the entity's constructed world- the Umwelt- from its internal functional world- the Innenwelt, stressing the close relationship that exists between these two different but complementary realities.

If the Umwelt corresponds to the entity's particular meaningful world, the Innenwelt can be said to be defined by the set of internal state(s) that characterize an entity's physical condition at a given time. Conceived as inherently systemic the concept of Innenwelt is essential to understand the phenomenon of salience emergence in cognition, i.e., why for a cognitive agent specific environmental features seem to get more salience, at a given time, comparatively to others. In fact salience is determined by the organism's needs as reflected by the states of its Innenwelt. These states defining the priorities of salience emergence in what concerns environmental features' prominence will drive the life form's actions and patterns of behaviour.

On the other hand variations or changes in its Umwelt, caused by structural change and consequent variation of parameters in the environmental bubble, will inevitably affect the organism's internal states and its drives.

The fundamental concepts of semiosis and Umwelt, Innenwelt are at the core of the mathematical modelling proposed by Ferreira and Caldas (2012). This model aims to capture the dynamics inherent to the semiosis of cognition. Placing the fundamental "interpretative" process at the core of the model it highlights the essential role played by the meaning assigning process in the cognitive process and consequently in the production of intelligent behaviour.

Figure 2. Like soap bubbles, Umwelten crisscross, partially or totally overlap, merge.

The diagram depicted in Figure 3 aims to model the dynamics inherent to a single instance of cognition, capturing how the dialectics established between the agent's internal states and its interpretation of environmental influx determine not only its subsequent actions and patterns of behaviour but the set of its internal states.

In the diagram, *y* is a vector of dimension[6] ($N_y \times 1$), which is assumed to represent all the potential information present in the general environment.

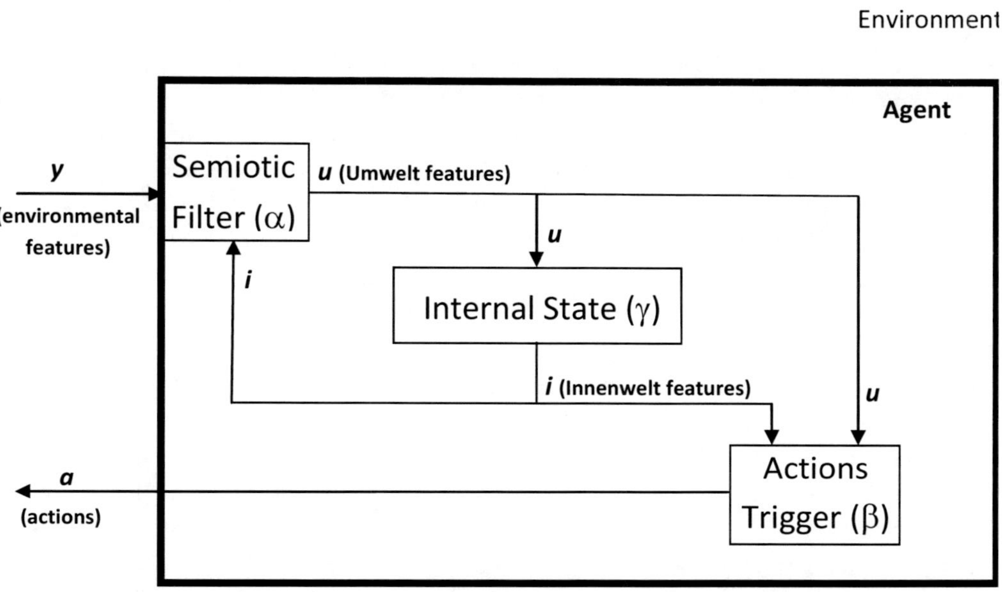

Figure 3. Modelling a single instance of cognition.

Acknowledging that only a specific set of environmental features will be perceived by the agent in the general environment and that even in this set the potentially meaningful features will get different prominence rates at different times and within different contexts, the agent's perception of its environment (Umwelt) is modelled through an ($N_u \times 1$) vector, *u*. This vector is created from the environmental features vector, *y*, through the application of a semiotic filter, *F*, whose characteristics will be dependent on the agent's internal state (Innenwelt) represented through an ($N_i \times 1$) vector, *i*.. The agent's particular view of the world- Umwelt- will then influence both its actions and consequent transition to a new internal state. This new internal state will, in turn, influence both the agent's actions and its semiotic filter, and, through it, its environmental perception. The vectors *u* (Umwelt) and *i* (Innenwelt) are, therefore, in a dialectic relationship and both of them are essential in determining the agent's actions.

It is assumed that there are N_a possible actions that can be executed by the agent and collect the respective probabilities of execution in a vector, *a*. These actions, if executed, will have a measurable effect on the environment, which could be observed by the agent in the next time step.

[6] In general ($N_l \times N_c$) indicates de dimensions of a matrix, N_l being the number of its rows and N_c that of its columns; thus, ($N \times 1$) represents an N-component vector in the form of a column matrix.

2. Umwelt Overlap and Ecological Sustainability

Uexkull's metaphor of the soap bubble shows how the individual self-world is in fact nested into a general environment where a plethora of other "soap bubbles"- individual worlds- coexist. These individual worlds are the Umwelten of all life forms that coexist at a given time. Ferreira and Caldas (2012) introduce the concept of Umwelt Overlap, viewing the general environment as populated by an infinity of individual Umwelten that crisscrossing, overlapping or even merging define common spheres where a collective semiosis takes place. The concept of Umwelt Overlap highlights how even slight variations in the parameters that rule an individual organism's world can drastically affect the universes and internal states and consequently the existence of a multitude of different life forms.

Giving as an example the phenomenon of nearly extinction of the *Lynx pardinus* in Portugal one can easily identify the series of structural changes introduced in distinct individual Umwelten responsible for this problem. The fact that man crossed initially secluded regions with huge motorways caused the extinction of vegetable species feeding the wild rabbit, the essential diet of the Lynx. Plus, the constant movement of cars through those regions determined the progressive killing and run away of the rabbit. Without its essential energy supplier the Lynx has progressively disappeared. The fact that human beings have introduced this structural change in their world totally unaware of the impact this change would have in the chain of interdependencies inherent to the overlap of three distinct worlds has nearly caused extinction of an already very rare species .

In his work, Uexküll claims for an essential epistemological turning-point in biology that in our opinion may have fundamental consequences for the identification of basic essential parameters responsible for the cohesion and sustainability of microcosms and ecological equilibrium. Uexküll assumes Kant's self-proclaimed Second Copernican Revolution as the foundation for the entirety of his own thought and makes this fact explicit in his introduction to "Theoretical Biology" (1926:xv)

> " When we admit that objects [...] owe their construction to a subject, we tread on firm and ancient ground, especially prepared by Kant to bear the edifice of the whole of natural science. [...] the task of biology consists in expanding in two directions the results of Kant's investigations:
> (1) by considering the part played by our body, and especially by our sense organs and central nervous system.
> (2) by studying the relations of other subjects (animals) to objects"

This fundamental epistemological turning-point involves a change of perspective whose consequences are twofold. On one hand it focuses on the essential role played by the physical architecture of life forms in the interaction with the surrounding environment and in the definition of a particular world; on the other hand this methodological change re-evaluates the status and standing point of the observer in scientific inquiry.

Uexküll has this notion when he writes (1934:15):

> "The Umwelt of any animal that we wish to investigate is only a section carved out of the environment which we spread around it_ and the environment is nothing but our own human

world. The first task of Umwelt research is to identify each animal's perceptual cues among all the stimuli in its environment and to build up the animal's specific world with them."

We could say that in his analysis Uexküll favours a horizontal perspective - a synchronic view[7]- that making a momentary pause in the existential timeline "freezes" phenomena, analysing the semiotic process in a snapshot, as it takes place in a particular fragment of space/time[8]. This does not mean that the evolutionary and developmental history of organisms and the evolutionary history of their respective environment aren't also present. In fact they always are, justifying it all. On this topic Ferreira (2011) points out:

> "Assuming a synchronic perspective, when analysing and describing the particular semiotic relations that take place inside a specific niche, one cannot forget that the cognitive entity has a history, past and present that constrains and determines the forms of interaction available. All cognitive agents are situated".

As Varela (1992) writes:

> " Ordinary life is necessarily one of situated agents [...] situatedness means that a cognitive entity has by definition a perspective. This means that it isn't related to its environment "objectively" that is independent of the system's location, heading, attitudes and history. Instead, it relates to it in relation to the perspective established by the constantly emerging properties of the agent itself and in terms of the role such running redefinition plays in the system's entire coherence".

In the analysis of semiotic processes essential to the correct understanding of what it means "ecological equilibrium" the synchronic and the diachronic perspectives are complementary. They are two virtual axes that crisscross in a single instance that is the entity or phenomenon under scope. It is the observer that defines and chooses one of the two distinct perspectives. When a "historical view" is assumed- Diachronic perspective- it is the entity's evolutionary and developmental history that is essentially under scope. When assuming an "existential view"- Synchonic perspective- the observer privileges the semiotic dynamics as it is, according to his own perception, actualised "here and now". In the synchronic perspective the observer analyses and describes phenomena in the apparently stabilized form corresponding to a defined moment in time, as the model by Ferreira and Caldas (2012), figure 3, proposes.

---

[7] The use of the distinction in scientific inquiry was first proposed by Saussure (1915). According to him language phenomena and consequently linguistic inquiry could be plotted on two axes: the *axis of simultaneities* and the *axis of successions*. The axis of simultaneities standing for the relations of coexisting things ... from which the intervention of time is excluded, while the axis of successions advances through time such that only one thing can be considered at a time. Fixed grammatical rules would occur on the axis of simultaneities, while their gradual evolution through linguistic history on the axis of successions.

According to Saussure, the synchronic and the diachronic views are two entirely distinct and complementary perspectives, though mutually excluding. Just as celery may be cut across the grain to display its cross section or may be cut with the grain to display its striations, all linguistic phenomena can be analysed according to a synchronic or a diachronic perspective. *Synchrony* emphasizes the axis of simultaneities that embraces grammar's relative permanence as a collection of syntactic rules and habits shared by individuals in the same language group, while *diachrony* emphasizes the "evolutionary phase" of language on the axis of selections such as occurs from one century to the next.

[8] We consider Darwin's approach as essentially diachronic, while Uexkull's is essentially synchronic

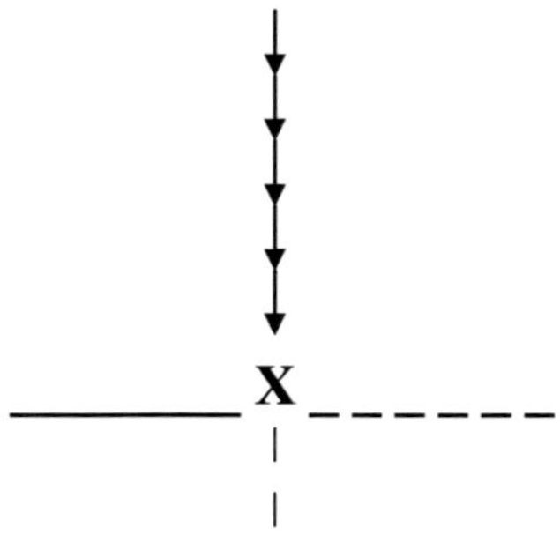

Figure 4. Diachrony and Synchrony.

In the diagram above, *X*, represents the living entity being observed. The arrows in the vertical axis represent the distinct phases in the evolutionary and developmental process associated to *X*. The horizontal axis situates *X,* "here and now", i.e. the phenomenon under scope is analysed by the observer in a framework that according to his/her own cognition corresponds to a stabilized form. The sequences not yet accomplished are represented by the dotted lines in both axes.

CONCLUSION

Semiosis is an embodied, embedded and always situated phenomenon. This means it involves a living entity endowed with a particular physical architecture interacting with the specific world it is immersed in, producing a dynamics that defines events taking place in a particular microcosm and perceived by a human observer as anchored in space/time.

The semiosis that is inherent to all forms of cognition gives consistency and coherence to the organic unit constituted by the agent and its meaningful world. The life form and its Umwelt are not two distinct entities but one structural and functional unit that must be viewed holistically, as any variation introduced in the parameters that rule out the functioning of any of these two parts will profoundly affect the coherence and functioning of the unit as a whole.

Uexkull's metaphor of the soap bubble illustrates how the individual self-world is in fact nested into a general environment where a plethora of other "soap bubbles"- individual worlds- coexist. The concept of Umwelt Overlap introduced by Ferreira and Caldas (2013) allows for the identification of the lattices of interdependencies established between life forms providing a key to the identification of essential conditions of ecological sustainability.

REFERENCES

Cassirer, E. (1996), *The philosophy of symbolic forms, vol.4,* The Metaphysics of Symbolic Forms, New York: Yale University Press.

Emmeche, C. (2000). Closure, Function, Emergence, Semiosis and Life: The Same Idea? *Reflections on the Concrete and the Abstract in Theoretical Biology.,* in Jerry L. R. Chandler.

Ferreira, M.I.A. (2007). *Ph.D. Thesis. On meaning: the phenomenon of individuation and the definition of a world view.* Faculty of Letters. University of Lisbon.

Ferreira, M. I. A. (2010), *On Meaning: a biosemiotic approach, Biosemiotics*, Springer.

Ferreira, M. I. A. (2011), *On Meaning: Individuation and Identity*, Cambridge Scholars Publishers, England.

Ferreira, M. I. A. (2011), I*nteractive Bodies: The Semiosis of Architectural Forms. Biosemiotics*, Springer.

Ferreira M.I.A. and Caldas M.G. (2012), *Modelling Artificial Cognition in Biosemiotic Terms, Biosemiotics*, Springer.

Ferreira M.I.A. and Caldas M.G. (2013) *Umwelt Overlap: An Aplication to Artificial Cognition , Biosemiotics*, Springer (forthcoming).

Maturana, H.R. (1978), Biology of language: The epistemology of reality, In G. Miller & E. Lenneberg (Eds.), *Psychology and biology of language and thought,* New York: Academic.

Merleau-Ponty, M. (1962). *Phenomenology of perception.* New York: Humanities.

Saussure, F.(1969). *Cours de Linguistique Génerale.* Charles Bally et Albert Sechehaye eds. Payot. Paris.

Uexkull, J.von, (1909), *Umwelt und Innenwelt der Tiere,* Springer.

Uexküll, J.von (1926). Theoretical Biology. Translated by D.L. Mackinnon. *International Library of Psychology Philosophy and Scientific Method. Harcourt,* Brace and Company, Inc. New York.

Uexküll, J.von (1934), *A Stroll Through the Worlds of Animals and Men - a picture book of invisible worlds, Instinctive Behaviour, the Development of a Modern Concept,* Claire H. Schiller translator and editor, International Universities Press, New York.

Varela, F.J. (1992), Autopoiesis and a Biology of Intentionality, *Proceedings from the Dublin Workshop on Autopoiesis and Perception,* essay 1 (available @ www.eeng.deu.ie/pub/autonomy/bmem9401).

In: Microcosms
Editor: Christopher C. Harris
ISBN: 978-1-62618-661-3

Commentary

A Brief History on Soil Microcosms as an Experimental Apparatus for Biogeochemical Research

***Kewei Yu*[1] *and Ronald D. DeLaune*[2]**
[1]Department of Biological and Environmental Sciences, Troy University, Troy, AL, US
[2]Department of Oceanography and Coastal Sciences, Louisiana State University, Baton Rouge, LA, US

Abstract

Most biogeochemical processes in soils involve changes in oxidation and reduction (redox) status, which can be characterized by redox potential (Eh). Determination of aeration status represented by Eh measurement can reasonably predict the stability of various compounds of biogeochemical interests. One experimental approach is to determine and to closely control the redox potential and pH at which various redox couples function. Such a redox potential – pH controlling device was originally designed half a century ago. Since then, it was named a "soil microcosm", because the incubated soil represents a miniature of the soil in the natural environment. Modifications have been made to suit specific research needs and to improve the performance of soil microcosms. The applications of soil microcosm techniques have significantly contributed to our understanding of biogeochemical processes among soil/sediment, plants, water and atmosphere, especially in wetland ecosystem.

1. Introduction

Following World War II, the rapid increase in global population resulted in unprecedented social, economic and environmental challenges. One of the most important challenges was food production for world population growth to meet basic needs and demands for rising living standards. Rice (*Oryza sativa*) is a staple food for a large part of the world population (FAO, 2003). As a part of the world-wide effort for the "Green Revolution" movement during the post WWII era, the International Rice Research Institute (IRRI) founded in1960 made a significant contribution for increasing global rice production. In addition to developing new rice varieties, understanding the soil chemistry in flooded rice fields to meet rice production became IRRI's major science endeavor. Early research in flooded rice fields (artificial wetland) made a major contribution in developing the foundation of today's wetland biogeochemistry, due to the similar hydrological settings as in natural wetlands. A large majority of the fundamental knowledge and principles in soil biogeochemistry were obtained from soil microcosm experiments.

2. Development of Soil Microcosm Apparatus

Soil aeration status can cause fundamental changes in soil chemistry. Unlike aerobic upland soils, flooded soils are devoid of oxygen supply from the atmosphere, and anaerobic conditions develop shortly after consumption of remaining oxygen by soil microbes. To understand the factors governing the growth of rice plants in flooded environments and for weed control and fertilization necessary for maximum yield, rice culture became a major driver in studies of biogeochemical processes in soils.

Most soil biogeochemical processes involve oxidation and reduction (redox) reactions. Oxidation reactions are a process of losing electrons, while reduction reactions are a process of gaining electrons. Oxidation reactions must couple with reduction reactions, which are called redox coupled reactions, to maintain the system electrically neutral. At the end a reductant (electron donor) is oxidized, while an oxidant (electron acceptor) is reduced in a coupled redox reaction. Soil oxidation–reduction status can be conveniently determined by redox potential (Eh). It has been observed that pH changes are associated with redox potentials when soils experience hydrological fluctuations. Due to the presence of oxygen and water, soil redox potentials are normally in a range from +700 mV (under fully drained conditions) to -300 mV (under prolonged flooding conditions) at near neutral conditions (Bohn, 1971; Ponnamperuma, 1972).

Simultaneous measurement of both redox potential and pH in soil solutions were conducted at IRRI to study soil chemistry for rice production (IRRI, 1964). It was recognized that soil redox potential and pH are the two dominant factors controlling the stability of various compounds of biogeochemical interests. However, under the same hydrological conditions, soil redox potential and pH were found largely different due to the difference in soil chemical characteristics, such as organic matter and various redox active inorganic components. Thus there was a need to determine and to closely control the redox potential and pH at which various redox couples function. Such a redox potential – pH controlling device was originally designed several decades ago (Patrick, 1966; Patrick et al., 1973). Since then, it was named a "soil microcosm", because the incubated soil represents biogeochemical conditions of the soil in the natural environment.

Typically, a soil microcosm involves redox potential and pH measurement in a homogenous soil suspension (Figure 1). Redox potential control (± 10 mV) can be achieved by adding either air (to raise Eh) or inert gases such as nitrogen or helium (to lower Eh) through a gas pump regulation system controlled by a meter relay. Most of biological and chemical processes tend to decrease the soil redox potential. When the designated redox potential is reached, the solenoid value is activated by the meter relay, allowing oxygen to enter the system to prevent the redox potential from decreasing below the designated value. In a rare case with a system where the redox potential tends to increase, redox potential control can be obtained using hydrogen-nitrogen mixtures. A similar system can be used for controlling pH (± 0.05 pH unit) in the soil suspension where pH electrode serves as the sensor. Acid or alkali is pumped into the system automatically through a solenoid valve to maintain the pH at the designated values. The system can be designed so that the amount of acid or alkali required is measured.

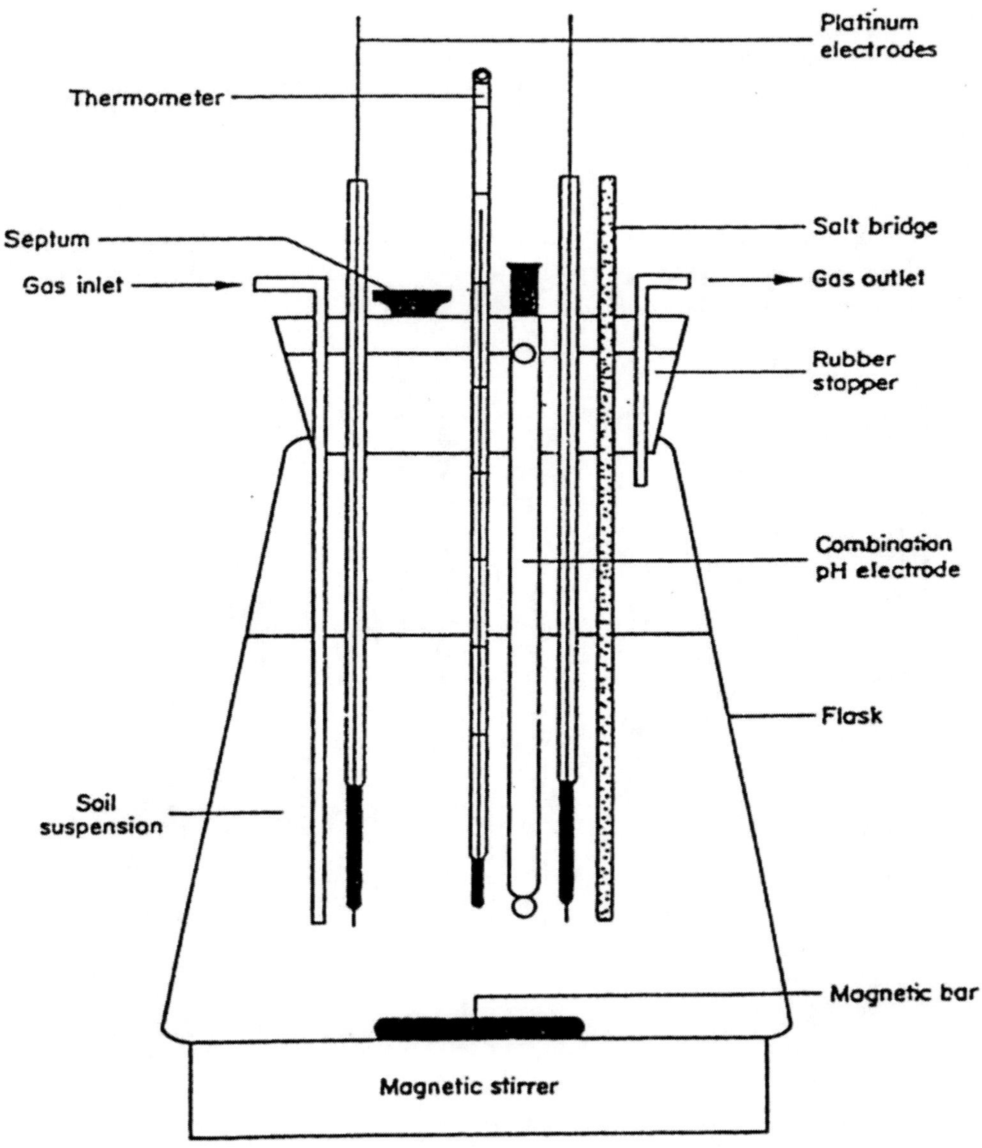

Figure 1. Schematic of a soil redox potential and pH controlling device.

In such a soil microcosm, samples can be taken as soil solution, soil slurry, and gas from the headspace when needed. Gases in the headspace produced by the soil microbial processes can be flushed out by using an inert gas, and can also be accumulated if the flush stops.

3. Modification of Microcosm for Special Needs

To improve the performance of the above soil suspension microcosm, several modifications can be made to suit specific research needs (Yu and Rinklebe, 2011). (1) To replace the rubber stopper with a screw cap where various openings can be prepared for electrodes and tubing through fitting unions. This helps the removal of the electrodes and tubing when needed. (2) To replace the magnetic stirrer bar with an overhead stirrer can make stirring the soil suspension more successful, but to maintain air tight for the system can be a challenge. (3) To replace the pH and redox meter with a data logger can significantly improve the frequency of data acquisition. (4) To provide temperature control for the microcosm can study the effect of temperature on biogeochemical processes. This can be done by using a heating blanket surrounding the incubation flask or by using a heating plate placed at the bottom of flask to reach above ambient temperature. To reach sub-ambient temperature (as well as above ambient temperature), a double-hull incubation vessel can be used where water of desired temperature can be pumped between the two layers of the vessel. (5) To install an automated gas sampling and analysis system allows the researcher to monitor the dynamics of gas concentration in a microcosm (Figure 2).

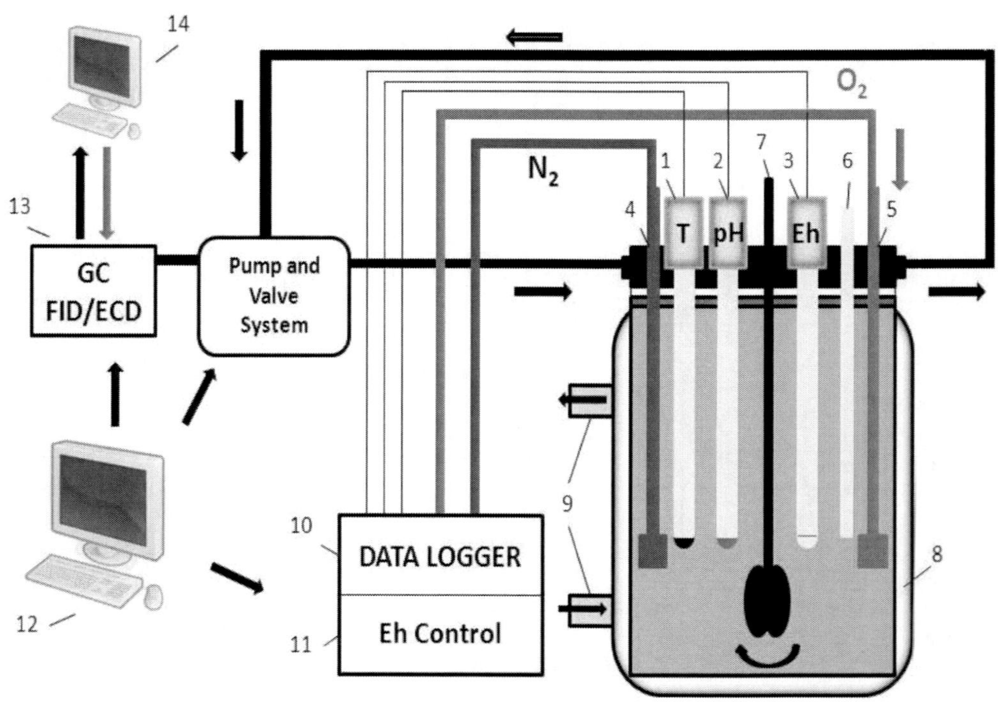

Figure 2. A fully automated soil microcosm with redox, pH, temperature control and gas analysis. (1) thermometer; (2) pH electrode; (3) redox potential (Eh) electrode; (4) dispersion tube for N_2; (5) dispersion tube for O_2; (6) sampling tube; (7) stirrer; (8) double-hull vessel; (9) temperature control by a thermostat and water circulation; (10) data logger for Eh, pH, and temperature; (11) automatic redox regulation by N_2 and O_2 valves; (12) control computer for data logger, pump, and valve system (gas sampling), and gas chromograph; (13) gas chromatograph (GC) with flame ionization detector// electron capture detector for trace gas measurements; (14) computer for GC control and GC data storage.

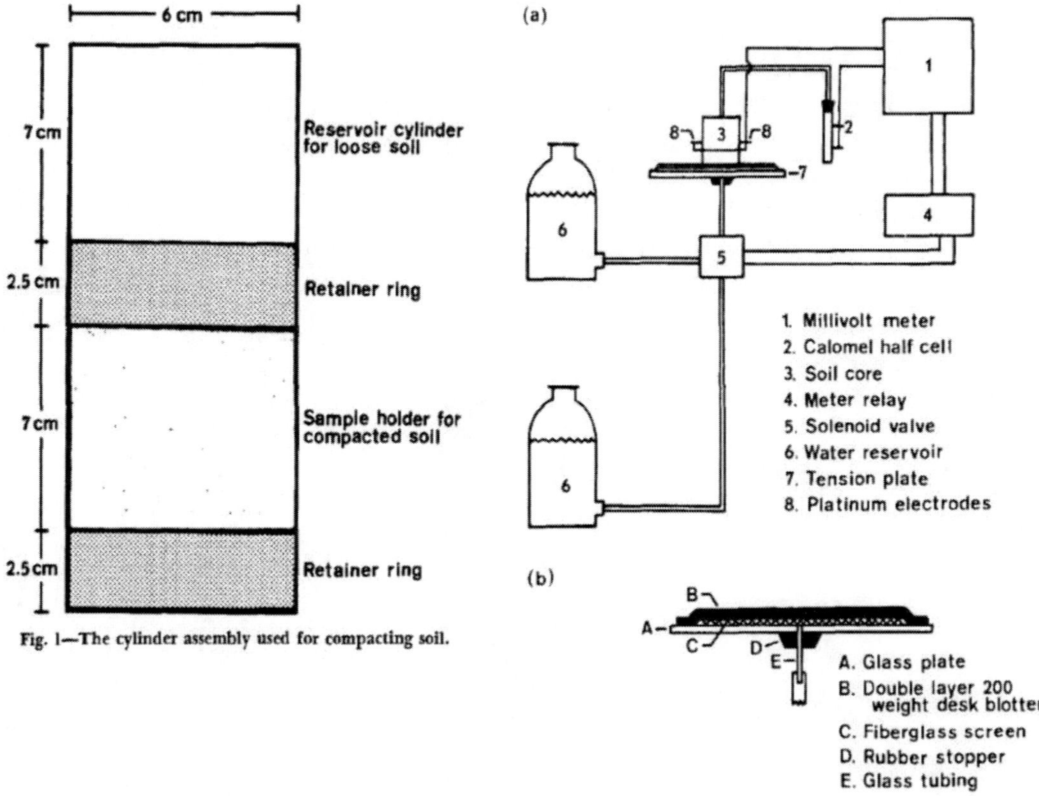

Figure 3. Schematic setup of a soil core microcosm.

A packed soil core can be controlled to reach a designated redox potential (± 20 mV). This is one step closer than a stirred soil suspension to a physically undisturbed soil (Patrick and Henderson, 1981). The principle of the soil core microcosm is to control the redox potential in soil cores by regulating the water tension. At a high water tension, the soil pore space is drained to let air enter to raise the redox potential. At a low tension, the soil pore space is filled up with water to lower the redox potential. A soil core can be prepared by packing the soil into a brass cylinder fitted with platinum electrodes on the side. The soil core is then placed on a moisture tension plate connected with two water reservoirs of different heights. Water tension control (redox control) can be reached by a two-way solenoid valve that is automatically controlled by a redox meter relay to connect to either of the two water reservoirs (Figure 3). Since rapid water capillary flow is essential for the soil redox potential control, this soil core microcosm can be readily applied only to coarse-textured soil. For a medium-textured and fine-textured soil, mixing the soil with sand is needed before packing the soil core. At the end of the study the soil core can be sectioned horizontally into slices for sample profile analysis.

A soil-plant microcosm was also developed to simulate the interactions between soils and plants in nature (Reddy et al., 1976; Tolley et al., 1986). The principle is to transplant plant seedlings into a redox-pH controlled soil suspension. In this microcosm, a desiccator base is used as an incubation vessel for the soil suspension, and a plexiglass plate is used as the cover for the vessel sealed with black plastic rubber or black silicone rubber sealant to maintain air tight. The soil suspension can be controlled for a target redox potential and pH as a regular soil microcosm (Figure 1). Various openings can be prepared on the plexiglass plate for the tubing, electrodes and plants to get through. Plant seedlings can be sealed to the plexiglass plate with a soft gum material. To study the gas exchange between plant and atmosphere, a transparent closed chamber can be placed on top of the plexiglass plate. Gas samples can be taken from the headspace of the gas collection chamber through a rubber septum (Figure 4). Soil chemistry analysis can be conducted as in a regular soil suspension microcosm (Figure 1). At end of the incubation, plant samples can be taken for nutrients and growth analysis.

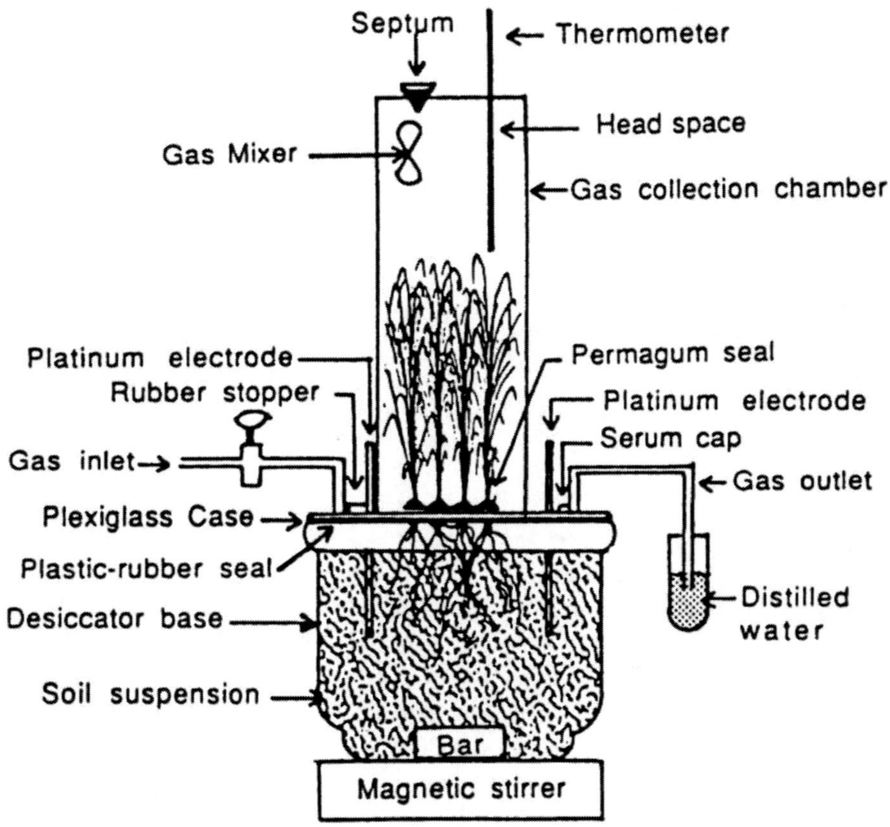

Figure 4. Schematic sketch of a soil-plant microcosm with gas chamber.

4. Application of Microcosm in Biogeochemical Research

It has long been observed that flooding can cause a pH increase in acidic soils but a pH decrease in calcareous soils (IRRI, 1963). Early research for rice production found significant accumulation of ammonium but rapid loss of nitrate in flooded soils. Flooding increased the content of soluble phosphorus and enhanced phosphorus uptake by the rice plants. Meanwhile toxic levels of heavy metals were also observed in flooded rice fields (IRRI, 1964).

The development of microcosm techniques significantly improves our understanding in soil biogeochemical processes of biological, ecological and environmental interests. The critical redox conditions for the reductions of major soil oxidants upon flooding are summarized in Figure 5 (Patrick and Jugsujinda, 1992; Yu et al., 2007). Two aerobic/anaerobic interfaces exist in flooded rice fields (1) flooded soil surface layer maintained by O_2 dissolved in the standing water, and (2) plant rhizosphere maintained by O_2 diffusing through the rice plant (Reddy et al., 1989). In addition to flooding and drainage soil management practices, these aerobic/anaerobic interfaces play an important role in the physiology and nutrition of rice plant. Lack of aerobic conditions in the flooded rice soils limits microbial nitrification activity, while dominant moderately reducing conditions favor strong denitrification activity resulting in rapid nitrate loss. Significant phosphorus release upon the flooding of rice fields has been associated with the soil iron reduction (Figure 5). Accumulation of carbon dioxide in the flooded soil may help to reduce the mobility of heavy metals and attenuate their toxic effects on rice plants. Similar results have been observed in natural wetlands.

Understanding various soil redox processes under different hydrological conditions makes fairly reliable prediction on the stability of redox active compounds in soil/sediment, water and atmosphere. This is not only important for fertilization and irrigation management for agricultural production, but also important for water quality in surface and ground water, as well as in marine environment (Reddy and Patrick, 1977; Masscheleyn et al., 1991). In recent years, global climate change becomes the most important environmental challenge where interdisciplinary and multidisciplinary research is urgently needed. Production of greenhouse gases and carbon sequestration in soils contributes to the long term sustainability of soil-plant ecosystem. Critical redox conditions have been identified for major greenhouse gases in terms of global warming potential by using soil microcosm technique (Masscheleyn et al., 1993; Yu and Patrick, 2004).

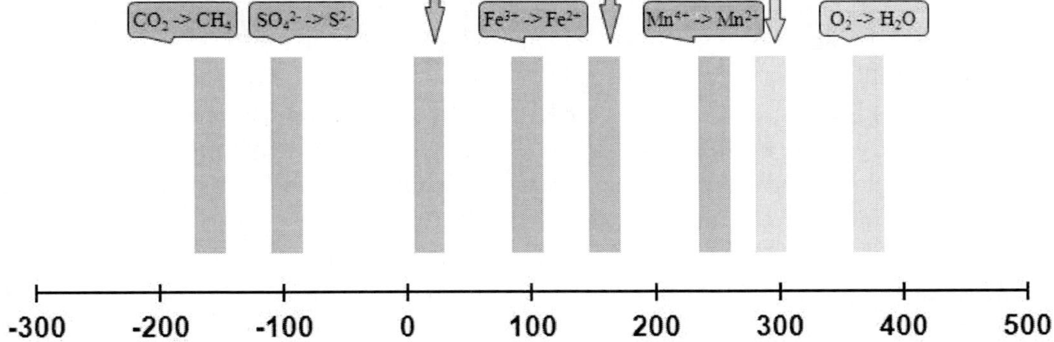

Figure 5. Critical redox potential conditions for major biogeochemical process in soils.

Advancement and modifications of microcosms makes possible a wide range of application of these techniques in ecological and environmental research, which contribute to understanding functions and services provided by the real world natural ecosystem.

REFERENCES

Bohn, H. L. (1971). Redox potentials. *Soil Science*, 112, 39-45.

Food and Agriculture Organization (FAO) of the United Nations (2003). Sustainable rice production for food security". *http://www.fao.org/DOCREP/006/Y4751E/y4751e00.htm.*

Masscheleyn, P. H., DeLaune, R. D. and Patrick Jr., W. H. (1993). Methane and nitrous oxide emissions from laboratory measurements of rice soil suspension: effect of soil oxidation-reduction status. *Chemosphere*, 26, 251-260.

Masscheleyn, P. H., DeLaune, R. D. and Patrick Jr., W. H. (1991). Arsenic and selenium chemistry as affected by sediment redox potential and pH. *Journal of Environmental Quality* 20, 522-527.

Patrick Jr., W. H. (1966). Apparatus for controlling the oxidation–reduction potential of waterlogged soils. *Nature*, 212, 278–1279.

Patrick Jr., W. H., Williams, B. G. and Moraghan, J. T. (1973). A simple system for controlling redox potentials and pH in soil suspensions. *Soil Science Society of American Proceedings*, 37, 331-332.

Patrick Jr., W. H. and Jugsujinda, A. (1992). Sequential reduction and oxidation of inorganic nitrogen, manganese, and iron in flooded soil. *Soil Science Society of American Journal*, 56, 1071–1073.

Patrick Jr., W. H. and Henderson, R. E. (1981). A method for controlling redox potential in packed soil cores. *Soil Science Society of American Journal*, 45, 35-38.

Ponnamperuma, F. N. (1972). Chemistry of submerged soils. *Advanced Agronomy*, 24, 29–96.

Reddy, K. R., Patrick Jr., W. H. and Lindau C. W. (1989). Nitrification-denitrification at the plant root-sediment interface in wetlands. *Limnology and Oceanography*, 34, 1004-1013.

Reddy, C. N., Jugsujinda, A. and Patrick Jr., W. H. (1976). System for growing plants under controlled redox potential-pH conditions. *Agronomy Journal*, 68, 987-989.

Reddy, C. N. and Patrick Jr., W. H. (1977). Effect of redox potential and pH on uptake of cadmium and lead by rice plants. *Journal of Environmental Quality*, 6, 259-262.

The International Rice Research Institute (IRRI), 1963. Annual Report, *http://irri.org/.*

The International Rice Research Institute (IRRI), 1964. Annual Report, *http://irri.org/.*

Tolley, M. D., DeLaune R. D. and Patrick Jr., W. H. (1986). The effect of sediment redox potential and soil acidity on nitrogen uptake, anaerobic root respiration, and growth on rice (*Oryza sativa*). *Plant and Soil*, 93, 323-331.

Yu, K. W. and Rinklebe, J. (2011). Advancement in soil microcosm apparatus for biogeochemical research. *Ecological Engineering*, 37, 2071-2075.

Yu, K. W., Böhme, F., Rinklebe, J., Neue, H. U. and DeLaune, R. D. (2007). Major biogeochemical processes in rice soils – a microcosm incubation from reducing to oxidizing conditions. *Soil Science Society of American Journal*, 71, 1406-1417.

Yu, K. W. and Patrick Jr., W. H. (2004). Redox window with minimum global warming potential contribution from rice soils. *Soil Science Society of American Journal*, 68, 2086-2091.

INDEX

A

B

C

D

E

F

G

H

I

L

M

Q

R